CO-AVZ-250

Successful Partnering

Successful Partnering

Fundamentals for Project Owners and Contractors

Henry J. Schultzel, CPA, CMC
Coopers & Lybrand L.L.P.

V. Paul Unruh, CPA
Bechtel Group, Inc.

John Wiley & Sons, Inc.
NEW YORK / CHICHESTER / BRISBANE / TORONTO / SINGAPORE

Library of Congress Cataloging-in-Publication Data:

Schultzel, Henry J., 1940–
     Successful partnering : fundamentals for project owners and contractors / Henry
J. Schultzel, V. Paul Unruh.
        p.   cm.
     Includes index.
     ISBN 0–471–11465–0 (alk. paper)
     1. Construction industry—Management.   2. Decision-making, Group.
3. Group problem solving.   4. Cooperation.   I. Unruh, V. Paul.
II. Title.
HD9715.A2S367   1996
692′.8′0684—dc20   95-21925

Printed in the United States of America

10 9 8 7 6 5 4 3 2 1

Acknowledgments

This work would not be complete without acknowledging the many people who have contributed to its success, including my partners and colleagues at Coopers & Lybrand L.L.P. and our special clients and friends at Bechtel whose participation in this concept proved what partnering can achieve.

I would like to thank Richard Van Mell, Evandro Braz, and Martin E. Gilmore for their extensive assistance in helping me develop our practical partnering approaches.

My special thanks to William R. Brink, whose premier assistance in the research and coordination of this project made it possible.

HENRY J. SCHULTZEL

San Francisco
September 1995

I would like to acknowledge the many fine people both in Bechtel and in our clients' organizations who have in the past and are now working hard to make the partnering concept what it is. Their achievements have been dramatic and beneficial to the concept. To those many people whose names I have inadvertently omitted, I apologize for doing so. For the few that I name, there are many more who have been on partnering teams and are too numerous to name. A few people who have contributed much and whom I thank: John Oakland, Bill Jackson, Michael Cole, Bill Teiser, Steve Grabowski, Charlie Duffy, Mike Gomez, Peria

Regupathy, Mark Zeiger, Errol Rapp, Dick McIlhattan, John Duty, Mike Storey, John Weiser, Gary Zieroth, Gary Weidinger, John Cooper, and Chuck Falk.

My thanks also to Charles B. Golson who assisted me in the research and preparation of this material.

V. PAUL UNRUH

San Francisco
September 1995

Contents

Preface: What Can You Expect to Get from This Book?

As we will discuss throughout this book, partnering is not a legal entity issue. Partnering is a working style whereby a group of specific people on a one-to-one and group-to-group basis build a culture in which they work closely and openly with each other to accomplish the extraordinary.

Key to the partnering work style is open, honest, and straightforward sharing of ideas and concerns and a willingness to challenge any policy, tradition, or protocol that stands in the way of optimum achievement and seems unreasonable to the group. As such, partnering groups have little patience with hidden agendas, ego issues, or reluctance to challenge the system. Small operations have been openly and enthusiastically working this way since the beginning of time. It is a key aspect of how they survive and grow. Small businesses that do anything less usually either go out of business or do little more than barely survive.

We could look to highly successful small operations for examples of the partnering work style. But, quite honestly, this book has been written with the big players in mind. One view has it that the big players are small operators who have done well, grown their organizations, and "outgrown" the partnering work style. When they are small, organizations seem to instinctively focus on extraordinary accomplishment, with policies and protocol being an afterthought more than a forethought. Once they are big, organizations seem to inherently start their thinking from established policies, traditions, and protocol, and move forward from there to do the best they can. The transformation from small to large is invariably accompanied by a transformation in motivation that goes from partnering when small to bureaucracy when large. The difference in perspective and motivation is not great, but the effect on achievement is significant.

So what can you expect to get out of this book? In some ways a return to yesteryear—i.e., a flashback to a time when you were involved in smaller groups that were on fire with a team camaraderie aimed at doing the impossible. For some of us those flashbacks will be to school teams or clubs; some of us will harken back to a crisis we were involved in; and some of us will think back to an especially energized work team where something like a partnering work style was a unique once-in-a-lifetime experience. But more than flashbacks, we will be working hard through this book to help pave the way to making the partnering work style a functioning part of most any situation. We will be focusing on engineering and construction settings and applications, but transferring these materials to any group setting should come easily.

In this book you will be exposed to dozens of different views of partnering: what it is, what it is not, what it is like, what it is not like, along with many examples from ordinary life experiences that help explain the principles and practices of partnering and how to make it work for you. At some point you may find yourself saying, "Enough already!" but read on and hang in there. It may feel like an overdose on too many different views, but when you are involved in partnering you bump up against a fair amount of resistance and misunderstanding. Over and over you will find yourself explaining what partnering is and is not, what it does or does not do, and how it works. The more examples and views you can offer, the more likely you will be able to make believers out of those who do not understand and supporters out of those who might otherwise resist you.

Through this book we are also attempting to provide you with insights and understanding to help you investigate, debate, decide, sponsor, launch, lead, manage, and do partnering. We will also present much of the mechanics of partnering, but please take careful note that the real secret to partnering is not in the logic or the mechanics, it is in the inner spirit of the people involved.

We expect that our readers will be coming from many different directions on partnering. To best serve a wide range of perspectives and prepare you to cope with a wide range of situations, we will be presenting you with insights and tips from those who have tried and failed as well as those who have struggled and won.

Each situation is different and each group of participants is different, but through this book we hope that we can offer you sufficient material with which to successfully build a partnering effort.

We sincerely hope that what we present here will be informative and useful. However, please understand that the statements and views in this book are our personal views and are not necessarily those of Coopers & Lybrand LLP or Bechtel Corporation. Furthermore, this book is not intended to render legal, tax, or other professional advice. If such advice is required, the services of a competent professional should be sought.

Successful Partnering

1

Why All the Hoopla about Partnering?

Partnering is not just good concepts, good engineering, good contracts, good contract management, good value engineering, good team building, and good project controls. Having and doing well in all those things can get you 90 percent of the way to making a project as good as it can be. In times gone by, 90 percent of the way to the best possible was about as good as anyone got or needed to get. But, in today's world, 90 percent will barely keep you in business. In today's world we must work our way above the 90-percent level just to stay afloat. So something more than good contracts and good teaming is needed if we are to work into that last 10 percent to make projects as good as they can possibly be, and that something is partnering.

This is a book about what works and does not work in the world of partnering. We have challenged ourselves each step of the way to answer for you the who, what, when, where, why, and how. We have strolled into the executive suites and climbed down into the trenches to collect material and ideas to provide insights and tips on what to do, how to do it, what to say, how to say it, how to know if all is well, and what to do when you find yourself falling short of the mark. We also share with you tips on what *not* to do and what *not* to say, equally important lessons in the world of partnering.

Those of you with experience in partnering, good or bad, will understand why this book is needed, and I am sure you will appreciate the value of what we are making available to you and your partners. For those who are presently on the sidelines of partnering, we are confident you will come to appreciate the no-nonsense insights and tips that you will find here. We will work hard to heighten your sensitivities to the important issues and build your confidence and skills in stepping up to the challenge.

SO WHAT'S THE BIG DEAL ABOUT PARTNERING?

Occasionally we have conversations with individuals who say they already have a good partnering arrangement. But the more we talk, the more we get the impression that they have good contracts and a good teaming arrangement, but to our way of thinking they do not as yet have good partnering. Good partnering is like icing on the cake. Icing is better than cake, but a good icing needs good cake as a foundation. In a like manner, partnering is better than good contracts and good teaming, but partnering needs good contracts and good teaming as its foundation.

If we are to engage in a meaningful discussion of partnering as a subject that rises above and beyond contracts and teaming, we must have a way to determine what parts of the cake we have and what we do not have. So, we will first turn our attention to understanding and testing for good contracts, good teaming, and good partnering.

TESTING FOR GOOD CONTRACTS

Good contracts are those documents, agreements, and understandings that each party considers to be appropriate for the situation (given the parties' intentions and information known at the time of contracting) and that, if executed well, will provide a framework for delivering the intended results such that each party to the contract will be satisfied with the outcome. Yes, it is a mouthful. But it provides a comprehensive single-sentence attestment that each party must be willing to sincerely agree to if they are to meet our test for good contracting. If all goes well the intended outcomes for the contracting will come to pass. And, if problems arise, the contracts make it quite clear how to address and resolve the problems.

Regardless of success or problems, with good contracts none of the parties will be in a position to justifiably cite a "bad deal" or flaws in the contract. It is of course true that some parties may cry foul no matter what the situation. A more objective and ethical test for good contracts in those cases might involve the opinion of a representative set of peers who are knowledgeable in the matters at hand and who would, after a review of the facts and situation, rule that the contracts meet our definition for good contracts (see Figure 1.1).

TESTING FOR GOOD TEAMING

We shall define good teaming as being present when the parties to the contract attest that they are comfortable that each party is focusing on what needs doing and how to proceed, and no parties are attempting to better their situation at the unreasonable expense of another party. In our definition we allow for normal assertive business intentions and practices, but we draw the line at working unethically against our contractual partners. Another test of good teaming is met when a group of partners indicates that it has established mutually beneficial working

Agreement			Good contracts are those documents, agreements and understandings that each party considers to be...
+2	+1	-1	
H	M	L	Appropriate for the situation, given the parties' intentions and information known at the time of contracting, and...
H	M	L	If executed well, the contracts will provide a framework for delivering the intended results...
H	M	L	Such that each party to the contract will be satisfied with the outcome.
Total Score ✍			GOOD CONTRACTS ≥ 3

Figure 1.1. Testing for good contracts.

relations demonstrating respect, concern, and commitment for achieving the overall purpose of the contracts (see Figure 1.2).

Good teaming suggests good working relationships among all parties, regardless of how "friendly" they might or might not feel toward each other. Teaming is good business, and it is what mature and respectable people do when they have common or codependent goals with each other. But in our definition we are not asking the team members to be like family, although good team members will often refer to their group as such.

Agreement			Good teaming is present when the parties to the contract attest that they are comfortable that...
+2	+1	-1	
H	M	L	Each party is focusing on what needs doing and...
H	M	L	How to proceed.
H	M	L	No parties are attempting to better their situation at the unreasonable expense of another party, and...
			The parties have established mutually beneficial working relations demonstrating:
H	M	L	Mutual respect,
H	M	L	Mutual concern, and
H	M	L	Commitment for achieving the overall purpose of the contract.
Total Score ✍			GOOD TEAMING ≥ 6

Figure 1.2. Testing for good teaming.

We are often asked if partnering is nothing more than exceptionally good teaming. Good partnering *starts* as good teaming, *builds* on good teaming, and when partnering goes awry it *falls back* on good teaming. But partnering is much more than just good teaming.

TESTING FOR GOOD PARTNERING

We will start this discussion with a summary test question for partnering (see Figure 1.3). We shall for the moment define good partnering as being present when the parties involved attest to having a camaraderie and a drive for mutual accomplishment that transcend the contracts at hand and involve genuinely, caring about each other's success, continually searching for ways to better everyone's situation, eagerly expressing concerns and ideas to help each other, and sincerely trusting that everyone in the group believes it is "one for all and all for one."

Hogwash, you might say; the engineering, procurement, and construction (EPC) industry does not work that way. That could have been my reply a few short years ago. We have been slugging our way through the real world for many years, and most of us have come to believe that any grinning fool in the real business world who thinks they are working with "a great bunch of people who are looking out for each other" can't be more than a few steps away from a rude awakening. Oh yes, we have all been in a situation or two where we *thought* we were in trustworthy company and could let our guards down. But we have learned our lessons, and we have learned them well. We have learned to say the words, smile the smile, play the game, and most of all we have learned to keep a sharp eye out for the "gotchas" that are lurking behind those phony smiles. We have been there, we have tried it, and we know better. We're no fools. . . .

Agreement			Good partnering is present when the
+2	+1	-1	parties involved attest to having a...
H	M	L	Camaraderie and a...
H	M	L	Drive for mutual accomplishment that transcend the contracts at hand, and...
H	M	L	Involves genuinely caring about each other's success,...
H	M	L	Continually searching for ways to better everyone's situation,...
H	M	L	Eagerly expressing concerns and ideas to help each other, and...
H	M	L	Sincerely trusting that everyone in the group believes it is "one for all, and all for one."
Total Score ✍			
			GOOD PARTNERING ≥ 6

Figure 1.3. Testing for good partnering.

But on second thought, most of us also can remember one time, a project way back when. It was one of those rare jobs where we had a lot of fun. The best part was that we all got along like we had been pals for years. We were always on the lookout for each other, helped each other with no questions asked, and saved each other's bacon time after time. We all shared a really special spark of pride and enthusiasm on that job. Thinking back, it is hard to believe we actually got that jazzed up over what we were doing. In hindsight, it was definitely a little crazy, but we really got into what we had to do. There just seemed to be a special energy in the air around our project. And, strangely enough, all the cards seemed to be stacked against us, too. But without any rhyme or reason, when we started forming our team something snapped into place. We all knuckled down, pitched in, and gave it all we had. We honestly believed no other group could have pulled it off as well as we did. We were unstoppable! It was as close to work heaven as you could hope to get.

As you can tell from the comments, we have been on the outside of partnering looking in, and on the inside looking out. And we have good news for you. In fact, we believe we have great news for you. In this book we will do our absolute best to share with you insights on how it is that a select few project teams pull together as "pals" and get energized to do the extraordinary. You will be glad to know that it is not as big a mystery as you might think. It also is not as complicated or weird as you might fear. However, it does take a solid understanding of the fundamentals, tips on how to work through and get comfortable with those fundamentals, a little soul-searching, a willingness to work through a few tough issues up front, and enough intestinal fortitude to struggle through two or three false starts in order to get your partnering launched. It is absolutely possible; it is absolutely doable. But it will be the rare person who simply leaps directly into great partnering. The rest of us face a step-by-step challenge, and we can only get to our destination one step at a time. So, please, stick with us on this. Again we promise you, it is doable, absolutely doable.

DOES PARTNERING REALLY PAY OFF?

Partnering opens the channels of communication more than ever before and optimizes the entire team's energy and creativity so as to accomplish the extraordinary (see Figure 1.4). Working together as "pals," the partnering team invariably comes up with a continuous stream of small but notable improvements, which add up to an impressive net gain by the end of the project. And hopefully, along the way, they also come up with two or three major improvements: sometimes cost savings, sometimes improved performance for cost for the final project, and often accelerated project benefits through completing and activating the project sooner than was expected. Good contracts and good teaming can certainly get the job done and lay the first 10 percent of success on the table, but it is partnering that gets us the next, best, and shiniest 10 percent. (Some readers will have already picked up on the problem of distributing the benefits of partnering across the

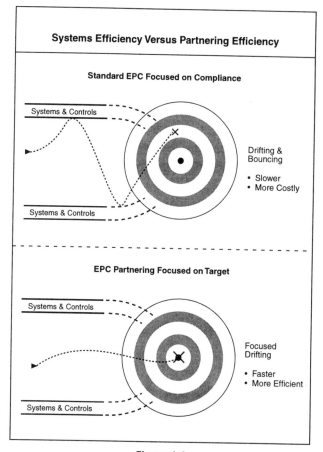

Figure 1.4.

various participants. We deal with that in a later section. So for now please assume that it is in everyone's best interests to continually improve the project.)

What Does Partnering Cost?

Perhaps the most amazing aspect of partnering is its own benefit-to-cost ratio. Without partnering, a number of the creative improvements never come to pass, so there is lost benefit. Without partnering, meetings take longer because of the usual positioning and CYA (cover your assets) moves. When difficulties arise, without partnering at least half the resolution time is spent jockeying around responsibility issues. The more difficult disputes often revert to contractual processes and legal posturing, and all too often are followed by the time and expense put into claims and litigation. Team members spend a fair amount of time thinking about whether or not they are giving up too much ground or being taken advantage of. And their worrying cripples and limits their creativity and enthusiasm, shows in their work, and it goes home with them and affects their family life. A significant amount of time is spent preparing ammunition for meetings and confrontations, and following up with letters and studies to analyze who did what, what they should have done, what they can do about it now, and doing whatever it is that they decide to do.

Without partnering, the best people and the best minds spend a disproportionate amount of time on business that is for the sake of pursuing and protecting their contract and protecting their backsides from all forms of imagined economic, image, and ego assaults. Partnering uses just about the same amount of time as would otherwise be wasted on what we have just discussed, but the trust of partnering refocuses that time on solving problems and being creative as a united force.

If the best people and the best minds are going to be working hard on something, why not spend their time working together on the end project's goals and values? The cost is essentially nothing, as they are going to spend the time and energy anyway. The benefit is that additional shining 10 percent of success.

Secondary Benefits

Secondary benefits from partnering can be as valuable as primary near-term economic benefits. Being recognized as a partnering player has an increasingly charismatic value. All else being equal, a partnering firm comes out ahead of a non-partnering firm. However, being recognized as a partnering firm can go beyond being a tiebreaker. It can provide a competitive advantage that can more than make up for other shortfalls, such as experience. Being recognized as a partnering firm can also have a perceived value in attracting new investors or competing for top talent. At the individual level, partnering executives may also be more successful landing desired board positions or speaking engagements. However, these secondary advantages to being recognized for partnering ability are dwarfed by true long-term benefits like being better prepared to handle future challenges and opportunities and fine-tuning operations for improved performance. Partnering benefits can also quickly spread to areas other than the original project that triggered the improvement.

A Wake-up Call for Doubters

Now, especially for those readers who may not yet be convinced to give partnering their best shot, let me offer a special wake-up call especially for you. I ask you to reflect on a different picture, a depressing and sad picture. It is a picture of readers simply breezing through this book without giving it serious thought. They are looking for a few good words, just enough "to be up on the latest fad." From the get-go they are convinced that partnering is hogwash, and they are looking for just enough of this silly "crap" to worm their way into the next contract.

They need just the right words to be convincing when they smile and emphasize how sincerely committed they are to partnering.

But at the end of the day those who don't give partnering serious thought will still be at the starting gate. They will believe that contracts and teaming do work, but partnering does not. And they will resist giving it more than lip service and a token effort. They choose to say the words, smile the smile, play the game, and most of all, keep a sharp eye out for the "gotchas" that are lurking around behind those phony smiles. They choose to never really try partnering. They choose to be a good team player, but never really step into the partnering game. So they choose to leave themselves in pretty good shape, but always just short of really getting into the spirit of partnering. They will never get good contracts and good teaming to the really energizing and rewarding center ring. They are forever wary of who might be taking advantage of them, who might be hiding something, and who is not as they seem. These players can have good projects, yes, and they can win the first 10 percent of success. But they won't be a part of the great projects we are going to be talking about in this book. They won't be a part of that highest 10 percent.

We must each choose for ourselves whether or not we give partnering our best shot. I am a realist, so I know that not every reader will give it their best. Too bad. But I am also an optimist, and that means that I also believe there are a lot of us ready and willing to go for the gold and get that last 10 percent out of a project. Thus, on that high note, we begin.

2

What Is Really Involved in Partnering?

In most cases partnering involves a major transformation of business attitudes and practices. It is a liberating process that strives to help the best that can be to realize their full potential.

Partnering champions the best ideas and practices in their emergence from traditional hang-ups and silliness. Partnering achieves extraordinary results because there is a largely untapped greatness in people, companies, and projects that wants to be set free. It is a greatness that for many years has been weighed down by layers of policies, procedures, protocols, and traditions. Partnering is a breath of fresh air that aims to remove the weights and set free the best in us.

TWO BASIC LEVELS OF PARTNERING

In this book we are going to view partnering on two levels. At the higher level we have the entity view—i.e., the situation of legal entities that are engaged in partnering. At the lower level we have a personal view—i.e., individuals who are personally engaged in partnering. Partnering can certainly exist at both levels at the same time, just as it can be missing from both levels. What may seem strange to some readers is that we hold that partnering can exist at the entity level and not at the individual level, and it can exist at the individual level and not the entity level.

In other sections of this book we will be presenting the mechanics of partnering. We will delve into how to plan it, do it, test it, and adapt it to a variety of situations. However, it isn't the mechanics of partnering that make it difficult or determine its success. The mechanics involve logical steps and approaches that generally make good sense to almost everyone in almost every situation. On the surface, nearly everyone can and will sign up to do partnering. What tends to separate the successful from the unsuccessful involves more philosophical and behavioral issues than it does mechanical issues. In this section we discuss some of the more important philosophical, cultural, psychological, and behavioral issues that are involved in partnering.

There are considerable similarities at both the entity and personal levels of partnering. What is generically involved in partnering is a camaraderie and an enthusiasm to do what it takes to accomplish the extraordinary above and beyond people just doing their jobs (see Figure 2.1). At the entity level we usually refer to such relationships as an alliance, which is an informal agreement that goes beyond legal contracts and requirements. In most cases alliances are based on non-binding agreements and involve an unusually open sharing of planning and resources to achieve otherwise-unavailable advantages in the market place. Most alliances involve linear cooperation as in integrated supplier relationships. However, an increasing number of alliances involve less linear blendings of technologies, capacities, contacts, and special resources that help the alliance participants to be more successful together than they would be operating independently.

At its highest level, partnering involves a new way of thinking and behaving. Participants in partnering enthusiastically help each other to avoid mistakes. If something gets in the way and hinders achievement, it is fair game to be changed to the better. Partnering's commitment is to extraordinary accomplishment. Excitement, creativity, enthusiasm, and zeal are the norm, and no partnering participant thinks of toning down or ridiculing high enthusiasm and energy.

To a large degree partnering is a very personal experience. It involves shedding our modern-day defensiveness and replacing it with a rebirth of sharing, trusting, and joining forces to do the extraordinary. It takes us back to that crazy school-spirit frenzy that many of us experienced in our younger years. In our youthful enthusiasm we tackled anything with all we had, and we expected to win every time. We made mistakes and laughed them off as learning, and usually pushed ahead more determined than ever. And if the establishment tried to protect its silly old ways, we took them on with a special zeal, knowing we could make the world a better place. We won some and we lost some, but we won more than we lost, and here we are.

As you read on, it might help to think about a time in your past when you were really excited about accomplishing something with a group, when you were really charged up and willing to go at it with all you had, a time when the setbacks just made you more determined, and you were energized with what you were doing, no matter how things were going. Maybe it was a youth club, a competitive team, a family challenge, a community project, or a public cause. Think about that experience. Think about what it felt like and what it meant to your self-worth and enthusiasm. Think what it would be like if that was what it was like in your day-to-day work. Those thoughts will get you well on your way to understanding what is involved in partnering in the business world.

When we find a group of individuals involved in a common business pursuit

PARTNERING LEVELS AND FACTORS		Agreement		
		+2	+1	-1
ENTITY LEVEL	Camaraderie	H	M	L
	Enthusiasm to do what it takes to accomplish the extraordinary above and beyond people just doing their jobs	H	M	L
	Open sharing of planning	H	M	L
	Open sharing of resources	H	M	L
ENTITY PARTNERING ≥ 4		Total Score ↙		
INTER-PERSONAL LEVEL	Camaraderie	H	M	L
	Enthusiasm to do what it takes to accomplish the extraordinary above and beyond people just doing their jobs	H	M	L
	No hidden agendas	H	M	L
	Honest	H	M	L
	Forthright	H	M	L
	Committed to best possible approaches...	H	M	L
	For best possible results.	H	M	L
INTERPERSONAL PARTNERING ≥ 8		Total Score ↙		

Figure 2.1. *Testing for partnering levels.*

who first and foremost are dedicated to accomplishing the extraordinary, we can say that they are in a partnering relationship. Partnering at the individual level means that the individuals have no hidden agendas, are honest and forthright in their dealings with each other, and are committed to finding and pursuing the best possible approaches for the best possible results. A partnering group will have a very low tolerance for policies, procedures, practices, protocols, behaviors, or traditions that get in their way and are not reasonably founded on ethical or legal grounds. When a partnering group faces an organizational or bureaucratic hurdle and can see a way through, under, around, or over it, they immediately shift into a "Why not?" mode. Hopefully, with much tact and skill, they confront counterproductive egos and traditions head-on. Their primary loyalty is to extraordinary accomplishments toward the end goal.

To draw a contrast, most work teams start with policies, procedures, practices, position descriptions, protocols, traditions, and objectives and set out to do the most they can from there. They want to accomplish the extraordinary, but their frame of reference is not the same as in partnering. Non-partnering groups generally accept, even honor, business as usual, and they make the most of it from a business-as-usual mindset. Most non-partnering groups are also far less energetic in breaking through business as usual, and when they face hurdles, they are much less likely to challenge them with the energy and determination that you find in partnering groups. Partnering is much more than just good teamwork. Good teamwork accomplishes what it is supposed to accomplish; good partnering accomplishes the absolute best it can, the extraordinary.

PARTNERING AT THE INDIVIDUAL LEVEL

Partnering at the individual level is not a department-to-department or a company-to-company thing. It is purely and simply a people-to-people thing. For you and me, partnering involves considerable soul-searching, sticking our necks out, opening up to new ways, and having the grit to stay the course until we accomplish what we have set out to accomplish.

Most of us were probably born with a partnering nature but have long since lost the knack. A few of us were born with a partnering nature and still have it, and a few of us were born with a nature that will probably never be comfortable with partnering. Whether or not you believe genetics plays a role in our basic nature matters little to our discussion here. It suffices to say that some people are ready, willing, and able to step into partnering relationships and will do very well. Some will need to face and resolve personal issues and build new skills to get comfortable and good at partnering. And some, for whatever reason, no matter how much they want it or how hard they try, will not do well in partnering and will either need to step aside or be pushed aside. We give everyone the benefit of the doubt and push ahead. Although not easily, we simply deal in a business-is-business fashion with those who fall short when that time comes. Our hope is that with this book in hand, many more people will come to understand partnering, will be able to understand what they need to do to succeed, and will have hands-on steps to follow to achieve partnering status.

Finding Our Starting Point   It stands to reason that the first step for us as individuals who will be involved in partnering is to figure out where we are on this spectrum. When we know where we are and we have a good understanding of where we want to be, then we can chart a course, launch our trip, track our progress, and stand a good chance of ending up where we want to be. So for the individual, what is involved in partnering?

When an individual joins partnering, they join a special group. For a partnering group, the nuts and bolts of partnering start with putting sufficient time in with each other to fully understand the group's situation, the group's challenge, where individual goals are alike, where individual goals are different, where individual goals are at odds with each other, doubts about how successful the project may be, concerns for each other's integrity or capability, and an honest leveling by each participant as to what they think they can and cannot offer to the project. And for us as individuals, the nuts and bolts of partnering start with whether or not we can sufficiently shed our defensiveness, egos, and hidden agendas to wholeheartedly join in with uncharacteristically open, honest, and forthright communications. Laying our cards, ambitions, and concerns on the table for others to see and react to sounds difficult, but it also sounds doable. However, when those who first get involved in partnering are asked what was most difficult in getting started, they often report that giving and receiving open and straightforward communication was their first and often greatest hurdle.

The Truth, the Whole Truth, and Nothing but the Truth As individuals pondering what is involved in partnering, we must start with the nature of the group we are joining. The partnering group has little tolerance and a very short fuse for hidden agendas, smoke screens, puffery, insincerity, or less than full commitment to the ending point of the project or mission at hand. Perhaps above all else, partnering involves individual and group honesty and forthrightness in the pursuit of the extraordinary. It is an enticing and invigorating opportunity, but it is wrapped in a logical but chilling commitment to "the truth, the whole truth, and nothing but the truth." But this partnering oath is taken in much more of an "all for one and one for all" sense, not a legal or adversarial sense.

Willingness to Blaze New Trails Those involved in partnering are more than a little fanatical about finding the best ways to accomplish the most impressive results as measured by their end product, even if they have to go out of their way to change the game or re-stack the deck. Partnering involves being willing to jump off old trails and blaze new ones.

If the best end result can be reached with existing practices and systems, so much the better for partnering. But if existing policies, procedures, systems, position descriptions, egos, power trips, politics, contracts, protocols, or standard practices get in the way of a partnering group, without hesitation the group will do its best to work around, over, under, or through the obstacle. Partnering involves a large dose of courage for acting on outdated, inhibiting, or illogical practices and behaviors.

Partnering teams certainly abide by legal and ethical guidelines. However, what they do not do is hold anything sacred from questioning or challenging. The further operations and functions are from ideal performance, the more critical and challenging the partnering team tends to be. Policies or practices that do not make a lot of sense or are not linked to specific legal or ethical requirements are generally fair game for attack if they get in the way. A frequent cry of the partnering group is "Why not?" Partnering involves shedding a "victim of the system" frame of mind and adopting a stance that is more assertive on behalf of the intended project or result. It is a little like quality circles, total quality management (TQM), continuous improvement, and business process reengineering (BPR) on the run in all directions and without bounds, except for the single-minded pursuit of the partnering team's mission.

Most work groups are either overworked or underworked. Those who are overworked can't seem to find the time to try and dream up better ways. Those who are underworked don't have any particular pressure to do more faster with less, so they also don't dream up better ways. But with partnering, dreaming up better ways to achieve more faster with less is an integral part of what they are and what they do. "Can we do this faster, better, or cheaper?" is an understood part of each conversation, planning session, coordination session, review session, and report. A driving force in partnering is doing the extraordinary, going beyond what was expected, going beyond what was thought possible. Partnering involves a very special camaraderie and enthusiasm that come from champion achievement, breaking records, and doing the impossible.

Partnering is simple. At its core it is simply the pursuit of good business and good business practices. But it is certainly not easy. For example, who would sincerely disagree with the following statements:

- Being honest and forthright works best.
- The more honest and forthright we are, the better off everyone will be.
- Being as creative and productive as possible can't help but put us ahead.
- Trying a lot of different well-thought-out things with a few failures is best; the good times and achievements will far outweigh momentary setbacks and individual embarrassments.
- Being honest with ourselves and open with others makes for a healthier culture and a less stressful life.
- We should all do our part to point out where company, department, and individual behaviors are counterproductive and need improving.
- Confronting and clearing out counterproductive protocol or ego trips is a responsibility we all share, and it can't help but improve our lot.

Good thoughts, right? But in our day-to-day business lives, taking one or more of these statements too much to heart and pushing it with any frequency will generally land us in hot water. Logic says that these statements are good and honorable. Reality teaches us that not all these ideas are welcome. Pushing them can backfire on us.

But there is another important truth that is also based in reality. Every infraction of these ideas costs us in time, resources, or stress. What is the price we pay for not pushing them? The answer in all cases is big bucks. Partnering involves learning how to push these ideas and ideals into the limelight in a way that gets the job done and allows us to survive. Partnering faces such issues with the drive and the heart to resolve them. But going to bat for these ideals and head to head against their counter forces isn't business as usual. It takes a lot of swimming upstream to make these ideals work. And swimming upstream takes careful planning, improved special against-the-current skills, and careful execution. Involved at the core of partnering are those rare skills that allow us to challenge the less-than-ideal and convince those we challenge that they will be better off for facing the situation and improving it.

PARTNERING AT THE ENTITY LEVEL

Partnering at the entity level involves two or more entities working above and beyond normal business practices to gain mutual advantage. The intent is to by-pass customary formalities in conducting business, such as integrating supplier relationships, or bridging customary barriers to working together, as when competing firms join forces to pursue new or difficult markets. In entity partnering or alliances, we are looking for those relationships that overlay and transcend legal entity issues. Generally speaking, for our discussions here it is not an alliance or entity partnering if the working arrangement is legally binding and conducted through a joint venture agreement, partnership agreement, or business contracts. It is only entity partnering to the extent that the participants work to mutual advantage above and beyond legal requirements.

Pursuit of Synergistic Benefit   At the core of an alliance is the pursuit of synergistic benefit, which involves a voluntary sharing of planning and resources based on mutual trust and respect. In almost all cases, alliance agreements are drawn up, but they are not generally legally binding agreements. They are more in the line of letters of understanding and intent. They outline expected values and behaviors, not strict policies and procedures. They outline visions, not requirements. Albeit extremely flexible, alliances are limited to those activities and re-sources that can be expected to generate intended synergism for the participants. Alliances are rarely all-encompassing to the point of helping each other just for the sake of helping each other. They are formed to achieve specific business advancements and objectives according to a mutual vision.

As with individuals, what is involved in entity partnering is a camaraderie and an enthusiasm to do what it takes to accomplish the unusual. A number of new entity behaviors are involved. The usual mistrust between entities is replaced with a willingness to openly share information and ideas. The usual tendency to hold back is replaced by a willingness to chip in wherever it makes sense. The usual tracking of each other and pointing out where the other party isn't measuring up is replaced by a joining of effort to help keep each other on track. The usual splitting of the economics according to contractual guidelines is replaced by a sense of fairness in reshuffling the costs and benefits out of respect for each other's situation and contribution.

Discipline Must Transcend Contractual Guidelines   Following and supporting the alliance isn't a legal issue. However, it does involve discipline. Individuals whose behavior is otherwise legal and ethical but which is detrimental to the alliance will generally find themselves suffering consequences little different than those suffered from infractions of ethical or legal guidelines. The alliance is serious business between the participants and is generally managed as if it were a legal and binding contract. For most alliances there are planning, execution, and review meetings that are essentially the same as those held for contracts, joint ventures, and partnerships. Expectations are made clear, tracking is evident in day-to-day communications, reports are prepared, and regular sessions are held to determine how things are going and what can be done to improve them. There is invariably a bottom-line focus by each participant to weigh the benefits and costs of the alliance, but the time line of the weighing is usually much longer than for customary business.

From time to time alliances are created to meet the participants' short-term needs, but most alliances are created to address longer-term strategic goals. This shift in time line from tracking on a near-term basis to a long-term basis is good news and bad news. It takes the pressure off evaluating each move and each effort and allows talented, well-intentioned, and knowledgeable people to work closely together to accomplish what is in everyone's best interests. But because the measuring is over the long term, day-to-day progress is often made on faith that the long-term goals will be reached. For alliances, long-term faith is a must. But alliances still need a reality check from time to time to ensure that the participants aren't just chasing a pipe dream or aren't in the end being taken advantage of by their fellow participants.

Alliances for the Long Haul   Honest and forthright sharing of information and ideas is at the heart of entity partnering. Within the scope of the alliance, participants' staffs behave essentially as if they were members of the same legal entity. However, the sharing stops short of crossing regulatory, legal, or ethical boundaries; and following sound business prudence, proprietary technologies and intellectual property may remain out of bounds.

As with individual partnering, entity partnering involves a commitment to finding and pursuing the best possible approaches for the best possible results. At the entity level it becomes a company-to-company and a department-to-department working relationship challenge. It involves clear communication from management to the staffs of all participants that alliance behaviors are not only okay, they are required. Less than enthusiastic support must be met with prompt identification and resolution appropriate to the best interests of the alliance's objectives. It involves a management commitment to manage and enforce requirements that are not exactly requirements.

INDIVIDUAL PARTNERING WITHIN ENTITY PARTNERING

Individual partnering is not a required practice for an alliance. Without the benefit of individual partnering, alliance business entities can work well together, fulfill their understandings of what should or should not be done to support the alliance, and meet their goals. What is involved in individual partnering within entity partnering is simply an extension of both concepts.

For example, if individual partnering is not present and an entity partnering situation is working well, then the alliance is providing the intended synergism for the participants. The participating individuals are sharing information and ideas and are looking out for each other's best interests within the alliance scope. But if individual partnering is not present, then their general mindset is to work within the intent of the alliance while honoring their organization's and employer's policies, procedures, practices, position descriptions, protocols, and traditions. They tend to think first of what is allowed or not allowed, and then do the best they can from there. They will accomplish the most that they can given their situation. It needs to be said that even without individual partnering in place, successful pursuit of a well-designed alliance is a powerful business advantage. However, adding effective individual partnering can leverage the alliance success to much greater heights.

What is involved in bringing individual partnering to the alliance is simply moving the alliance down to a personal level for those individuals who are interfacing between the alliance participants. It involves allowing them to think first and foremost of the mission and how best to accomplish the extraordinary. As we have already said, individual partnering involves allowing those involved to blaze new trails. If the best end result for the alliance can be reached with existing practices and systems, so much the better. But if existing policies, procedures, systems, position descriptions, egos, power trips, politics, contracts, protocols, or standard practices get in the way of the alliance partnering group, the group will need adequate support to do its best to work around, over, under, or through obstacles.

When individual-level partnering turns to challenging participating business entities in the pursuit of normal entity-to-entity business, tolerating and supporting their persistence is a monumental challenge for the management teams of the different companies. When individual-level partnering starts challenging alliance participants who don't even have a binding contract to fall back on for facing and resolving challenges, then management support for the challenges is treading on really thin ice. The goals are still valid, but the immediacy and relevancy are much more distant. This usually requires the individual partnering group to be more selective and savvy in their challenges, and management to be more sophisticated in their handling of the inter-company challenges.

At the end of the day, perhaps what is most involved in developing and stabilizing partnering is the intestinal fortitude to weather the storms that lie in the way. Establishing and maintaining partnering requires abundant stamina at both individual and entity levels. Pushing you along the partnering path are hopes of benefits to be and encouragement from those who keep the faith. But strong in your face are many headwinds pushing against you. Policies push against you. Tradition pushes against you. Protocol pushes against you. Contracts push against you. Group thinking pushes against you. Powerful people push against you (and some of them are the same people who have gone "on record" about how dedicated they are to partnering). The rewards are well worth the effort, but it takes a large dose of faith and courage to stay the course.

WHAT DOES AN INDIVIDUAL NEED TO DO TO GET READY FOR PARTNERING?

The most essential preparation for partnering is gaining essential knowledge and lore about what partnering is and how it works. Those entering partnering for the first time must gain a solid understanding of the many processes involved and the various forces that work for and against success. Those aspiring to partnering without this essential knowledge will at best find themselves going through the motions, using the materials, and wondering why partnering isn't turning out to be as golden as it should be. Their disappointment will not be a result of partnering not working, but rather a result of their poor preparation. The knowledge needed can be gained by reading and rereading this book.

This book will go a long way in helping you complete your preparation with partnering approaches and materials. After you begin your project and your partnering effort, you may have little time for developing materials and staying ahead of the game. There simply won't be enough time to figure it out as you go. You will need reading materials for your partnering group, group and self-assessment forms, meeting agendas, discussion guides, group exercises, discussion and decision tools, and a well-engineered planning, scheduling, and tracking system. Much of what you will need you will find here, or at least referenced here. Thinking through what you face and who you are facing it with, you should be able to put together all that you need to launch and sustain good partnering.

FACILITATOR'S ROLE

In most cases, groups that have been the most successful with partnering have used a facilitator to provide the knowledge, materials, training, support and discipline that partnering requires. It is not so much that the group needs an outsider to nurse them along, although that is a part of it, but more importantly it is best to have someone championing partnering who is not subject to the immediate demands and pressures of the project at hand. Having a facilitator also provides an independent resource to anyone who needs a little help or a little nudge. If one or more of the group members are the partnering promoters, they must split their time between partnering roles and project roles, and partnering characteristically comes out on the short end of the stick. At times it may also be disconcerting when a group member switches back and forth between content (their project role) and context (their partnering role).

Along with each individual doing their part, someone must spend the time and energy to promote and support partnering. It seems to be most reasonable to let that someone be an independent facilitator. The facilitator can come from any one of the participating firms or from outside the group. When possible the facilitator should be selected by the group. Three or four prospects should make presentations to the group covering their understanding of partnering and their experience. They should also discuss in detail the approaches they propose to use to introduce, support, and maintain partnering. The group should ask for specific examples of how each facilitator has handled other similar situations and should challenge the facilitator to work through a couple of scaled-down examples of dealing with difficult situations during the proposal presentation so as to see how the interaction goes. The group should then contact the facilitator's references and discuss the facilitator's strengths and weaknesses and consider any advice that the references might offer regarding a new assignment with the facilitator.

GETTING EXPECTATIONS IN LINE WITH REALITY

Sometimes people will expect partnering to produce miracles. Sometimes the hype about partnering contributes to unrealistic expectations. When the expectations for partnering are unrealistically high, the excessive pressure on partnering can be its undoing. Partnering does not presuppose that anything in particular can be done, other than extraordinary accomplishment under the circumstances at hand. When partnering falls short of expectations, it can generally handle the disappointment and keep on the path to do the best possible under the circumstances. But when partnering falls short of unreasonable expectations, it may also come under fire from those who think they can fix it, those who think they can put it back on track, and those who want to eliminate it and switch to something they think is better. Such situations are a matter of reality and something partnering simply has to deal with. Most cases dealing with unreasonable expectations are out of partnering's control. But if the situation arises because partnering has in fact not done a proper job of public relations and keeping key players informed, then it has failed.

Partnering must identify and deal with anything that might interfere with extraordinary accomplishment, and that includes dealing with image, public relations, and key executive relations, especially hostile relations. Partnering must be bold and creative, and that includes working effectively with hostile non-partnering players. Searching for and disarming those who are out to get or kill partnering is a part of the challenge. Ignoring or avoiding challenges that eventually derail partnering is not an acceptable path. Groups who forge ahead toward extraordinary accomplishment with blinders on will almost certainly be derailed at some point. To be successful, partnering requires a full heads-up and eyes-open posture.

ALTERNATIVE DISPUTE RESOLUTION AND PARTNERING

Some in the industry view partnering as primarily a working relationship built between contracting parties in order to minimize adversarial tendencies and ensure the highest level of cooperation. The focus is on open communication with an emphasis on addressing and resolving disputes without resorting to litigation. This general view of partnering has risen out of the early efforts by the U.S. Army Corps of Engineers to relieve projects from the substantial costs and delays involved in disputes, claims, and litigation. The Corps determined that the source of nearly all these problems came from a basic adversarial relationship between the contracting or regulating parties and a predominate focus on preparing rather than resolving disputes and claims. Their early efforts to resolve this situation through workshops and team-building steps have led to solutions for much more than just the dispute problems. Those early efforts have gradually matured into the more complete version of partnering that we are seeing today. Alternative dispute resolution is still an important aspect of partnering, but creativity, breakthroughs, and extraordinary accomplishment now steal the limelight.

PUTTING PARTNERING IN PERSPECTIVE

Because formal partnering efforts are a relatively new business endeavor, there is a tendency to tackle partnering as if it were a horrific new challenge. Individual preparation is often followed by a series of intense workshops, which are in turn followed by ongoing training and education to continually fine-tune partnering. Although we agree that there needs to be a launch plan, and we agree that there needs to be an ongoing planning, doing, checking, and adapting cycle, we want to emphasize to our readers that the basic intent of the launch effort is to put in place enough of a system that it can switch on to extraordinary accomplishment and become self directing and self managing. Having a system that works for the duration of the project is a start, but what we are really after is that moment when it takes on a life of its own. Only then does partnering truly begin.

Partnering does require a considerable amount of training and development. With a facilitator's or sponsor's help, once activated the group will decide on an ongoing basis if it needs to increase or decrease training. Not enough training and the group will fall short of its potential. Too much training and it will waste valuable time. Too little training is much more detrimental to extraordinary accomplishment than too much, so most groups should plan to err on the side of too much rather than too little. Experienced facilitators will have a good feel for how much training the group will need. Exactly when which training will be needed and to what depth depends on how well the group is doing and what challenges it is facing. The right amount of the right training at the right time is not something that can be planned too far into the future. Good facilitators work with the flow of the group just as the partnering group works with the flow of the project. Partnering is fast-track management from day one, and that includes considerations for training.

Partnering involves exceptionally open interpersonal and inter-entity communications and a willingness to challenge tradition and protocol in a relentless pursuit of extraordinary accomplishment.

On a day-by-day basis, free-flowing and essentially unrestricted communication of ideas and concerns is what is most involved at both the entity and individual levels of partnering. Take away open communication and you not only lose your partnering status, you lose any real chance for extraordinary performance. Take away the willingness to challenge tradition and protocol, and you take away the leverage to get the most from open communication. Have anything less than a relentless pursuit of extraordinary performance and you will likely achieve little more than good performance.

TRUST AND SHARING ARE CRITICAL

To be effective, partnering clearly involves and requires exceptionally high levels of trust and sharing. Unfortunately, by nature neither individuals nor organizations are very trusting or sharing. We have learned, and learned well, that while trusting is a great aspiration, trust infractions are so damaging that full trust is out of the question. Most individuals and organizations could be twice as trusting as they are and be much better off for doing so, but the consequences of even just a few trust setbacks are prohibitive or at least perceived to be prohibitive. We seem to assign treble damages to the setbacks, so we go out of our way to guard against them. High levels of trust are achievable, but we must be realistic and not berate ourselves for not achieving the ideal.

The same basic phenomenon applies to sharing. As individuals and organizations we could generally gain a great deal by openly sharing our talents and ideas, but there are enough of those who take advantage of us that we all have become at least somewhat guarded. Sharing too much at times can do us damage, and when it happens we assign treble damages to the setback. As with trusting, most of us could be twice as sharing as we are and be much better off for doing so, but the consequences of even just a few sharing backfires are prohibitive or at least perceived to be prohibitive. And so by nature we are generally not prone to openly trusting and sharing; in fact we are prone to staying alert and keeping our guard up.

Learning to soar with trusting and sharing means that we must build new skills and master our fears. It is a little like learning to ride a bike. Yes it is scary, yes we take a few spills in the beginning, but after a day or two of spills and a few bruises we suddenly find ourselves gliding along feeling as if we are on top of the world. Getting past the first few spills and fears opens the door to a whole new world of speed and fun.

How is it that some people never get past the spills, fears, and bruises, and others do? There are two basic reasons, and both apply to learning partnering just as much as they apply to learning to ride a bike. First, some learn to ride a bike because important people in their life in one way or another force them to do it. These people go on and on about how much fun the other kids are having, about how you are just as able as they are, and how they will run alongside of you and hold you up to keep you from falling over. Whether in a kind or tough way, these people who are showing an interest in you see to it that you learn to ride a bike. Some of us face that situation in learning partnering. Someone shows an interest in our succeeding and through support or threat sees to it that we make it through the initial stages. Almost without choice we will be pushed into and through open communication, trusting, and sharing, and after a few disappointments, embarrassments, and crashes we will find ourselves sailing along at a new level of extraordinary accomplishment. As with bike riding, all of a sudden we can do it and away we go.

The second reason some people learn to ride a bike is to satisfy a personal burning desire. It might be the need to impress someone very important to you or to achieve some necessary goal, like getting somewhere quicker or qualifying for a paper route to earn needed spending money. For whatever reason, and with no one else around, you get on the bike over and over, spill after spill, until you can ride and win the admiration you sought or gain the goal that was so important to you. Some people see partnering working, or at least come to a point where they believe partnering will gain them what they want, and they jump on board. They have a burning desire to succeed, so they plow through the stumbling and setbacks until they are off and running. As with bike riding, it doesn't matter what gets you through the learning stage, what matters is that you do get past the spills and get to the rewards.

The reason why open, straightforward communication is such an issue in partnering is that it is the only effective means to the intended end. It is like getting on the bike and shoving off. It's not what riding a bike is all about, but it's the only way to learn to ride a bike. Trust between partnering participants comes from communicating ideas and concerns and finding out that it's okay, i.e., communicating expectations of each other and finding out that your partnering participants are being truthful, honest, and honorable (even if they can't always actually deliver on what they say). Open communication is the mechanism to set the stage for trust, and it is the means to testing each other to see if the trust is well placed. When ulterior motives, hidden agendas, false promises, prejudices, subversiveness, and the like are dispelled through open communication, and follow-through and sincerity are left, then trust is well on the way. When, through open communication, you find that your ideas and concerns are respected and dealt with fairly, then your natural guard against sharing begins to drop. Open communication is essential to maximizing the talents and potential of the group, but it is also the means that is needed to get to the trust and sharing levels that determine how successful a partnering group can be.

3

A Word to the Wise

The writing is on the wall. Partnering is a popular item in the press, and it is a popular item in planning meetings. There is little doubt that as owner or contractor, no matter who you are or what you do in the engineering and construction industry, to be successful over the next few years you should get involved in partnering. You would do well to get involved sooner rather than later and get good at it, or get out of the way of those who are. We will each make our own choice. If you choose to get out of the way or are pushed out of the way, there is obviously still a place for you and many others like you outside of partnering. But the real success stories and champions of the industry will most likely be the best partnering participants. Those firms and individuals who aren't into partnering likely won't be in the big-time playoffs under the bright lights. The same old ways will produce the same old results, and only extraordinary ways will produce extraordinary results. Partnering is clearly the best extraordinary way of doing business available to us in the industry today.

So a word to the wise is to get involved, but another word to the wise is to go in with heads up and eyes wide open. Almost as many as rush forward into partnering will quickly slide backward into their old ways. Partnering makes a lot of sense and sounds reasonably doable. But in truth it is difficult, very difficult. The rush of entities and individuals signing up for partnering is only slightly greater than the rush of entities and individuals slipping back into old practices, back to the safe and well-traveled path of the known and the comfortable. So a word to the wise is to go in with what it takes to stay the course. Rushing in carelessly and slipping back to old ways involves embarrassment, lost trust, wasted time, and wasted resources. The wise will check it out thoroughly, plan well, and sign up to stay the course, not just take the leap.

If there isn't a great opportunity or huge challenge or an intense sense of urgency facing the partnering group, then partnering is usually hard to get off the ground. Champions don't just kick the ball around and come out on top. Champions see an intense struggle in front of them and have a burning desire to accomplish the extraordinary. More than anything else, they want the highest trophy in the land and all the glorious trappings that come with it. If your partnering group is facing such a struggle, partnering can catch on and do great things. If your partnering group is not facing such a struggle, your first challenge is to create one. Even in day-to-day business we can stir up a desire to be known as the best and visualize the trappings of business heroes. So another word to the wise as you begin this book is to be honest with yourself and your group about what you are setting out to do. If you are setting out to do good partnering, you are unlikely to succeed. If you are fired up to accomplish the extraordinary and have a clear and exciting vision of what you want to accomplish, then partnering can be just the ticket you need. Partnering is the means to an end, and the end must be clear, understood, and well worth the effort.

A special word to the wise for old hands of the industry: Old dogs behind closed doors will often confide that they think this partnering stuff is for sissies. It is clear to them that it's just a smoke screen for those who just don't know how to get things done. But by the time you finish this book I think you will agree that the truth is that essentially every situation can be improved by partnering. And an additional truth that will come to light is that you have to be seasoned, mature, and strong to do well in partnering. The "sissies" don't do much better in partnering than they do outside of partnering. However, it is true that within partnering the weak and insecure do have a better chance to grow faster and overcome their shortcomings.

Another word to the wise is that partnering is as much a culture and a style as it is a work process. It involves a significant step up in values and thinking and a special set of very mature behaviors. You start with the words and the mechanics and going through the motions. But partnering is not partnering until the individuals buy in and are fully energized in a new way of conducting business. An absolutely essential aspect of buy-in is every participant's willingness to relentlessly acknowledge anything that gets in the way of extraordinary performance. Especially important to this relentless pursuit is a willingness to be on the constant lookout for any aspects of the group or individuals that are causing or could cause a problem, and that includes a heavy dose of self-awareness and self-testing. Even the slightest flaws in individual or group participation need immediate straightforward attention, or partnering can be seriously crippled or derailed.

Partnering is also a process that must evolve group by group. You can't just mandate that it be done and it is so. It is a little like making it into the finals in a league. You can round up a team, explain all the rules, and lay out a series of winning plays. But it takes a lot of camaraderie and zeal for the team to start consistently winning games. Nearly all the teams in a league have the same basic set of players, coaches, and facilities. Their contracts, policies, and procedures look much more alike than different. But the winningest teams have a special aura about them. In an otherwise ordinary sport, there is something extraordinary about the teams that consistently win.

In some ways partnering is like that winning team aura. It is something extraordinary about a group of people who are involved in ordinary business. You plant the seeds of partnering, you provide the nutrients, water, and sun to set the stage, go through the motions and get things started, but you can't force it. Partnering has to grow within itself and become a living, breathing part of the group and its challenge. So we have another word or two for the wise: You can't "do" partnering. You can't just say "let's do partnering" and expect it to work out. You have to "become" partnering.

Right now partnering is still somewhat of a fad. It is a tool that we sometimes choose to use, sometimes not. And as with most fads, we tend to give it a whirl but are usually not very successful at it. So too often after a so-so try we drop the fad. However, in time we believe that partnering, unlike most other fads, will be at the heart of the way the truly successful movers and shakers in the industry manage and work together. More and more we will hear about the proven value of internal and external partnering. As it becomes more and more ingrained in what we do, the essence of it will show up as standard line items on surveys, performance reports, and management reports. For a few more years the majority of individuals may choose to not play or won't be able to make the transition from traditional business practice to partnering. But, another word to the wise, eventually partnering managers will fill the majority of key coordinating positions. Over the next 10 years we believe the scales will tip dramatically toward partnering styles and practices in the workplace, especially in the world of successful engineering and construction. The wise will study, learn, practice, and excel in partnering principles, values, and practices.

WHAT IS LIKELY TO GO WRONG WITH PARTNERING AND HOW BAD CAN IT BE?

We could not possibly count all the project teams that have signed up for partnering and have launched their campaigns. But we would not have too much trouble counting those that have succeeded. Why is that so? What goes wrong?

Partnering has a lot to do with personal values, interpersonal skills, and working by getting along with people to the point where we can trust each other to work for each other's best interests. As such, partnering is not terribly complicated, but it is deceptively difficult to achieve. Our behaviors, habits, and values do not change easily. In setting out to change these behaviors, numerous things can indicate that we are making progress, but one or two small missteps and we quickly draw back, deciding all too fast that it won't work.

In partnering, as with so many other things in life, success is often only a few more missteps away. So many quit just short of success. But some people are willing to hang in there through the next missteps, and they are the ones who succeed and become our heroes. In most fields, sales statistics show that success usually doesn't happen until after five, six, or even seven sales calls. Yet, most salespeople give up after the third or fourth call. The top salespeople are generally those who hang in there for just a couple more calls. This is not a secret; it is well known. And yet the saga of those that do and those that do not goes on. It seems to be a part of the human condition that a few will hang in there and succeed, and most will stop just short of success. The same phenomenon applies to learning a language, mastering new software, excelling in sports, or meeting most any other challenge. Most of us can do it, if we just hang in there long enough. And so it is with partnering.

Genuine attempts to get partnering launched most often fail because the participants do not hang in there long enough. They hit a few bumps early on, and they find themselves reverting back to their same old ways, gradually leaving partnering behind. It is human nature, but it can be overcome. Two or three of the lead participants in a partnering group must commit to each other to keep trying and working the partnering plan for at least six months. If they will not consider backing off or giving up until after the six months are up, they will almost always succeed. It often takes two or three months for partnering to start really settling in, and another month or two to become self-sustaining. Unless partnering efforts are planned and scheduled for at least six months right from the start, and unless that plan is followed no matter what, then achieving partnering will be in serious jeopardy.

As an aside, when partnering fails because the participants fall just short of success, it is a double injury. First, the project and the participants will not enjoy the benefits of partnering, and second, the participants will be much less likely to give partnering a chance in the future. If you are re-starting a group that has attempted but not succeeded with partnering in the past, you will want to spend time up front discussing those past experiences and why this experience is going to be different.

The second most likely cause of partnering failure lies in failing to learn, use, and enforce proper interpersonal communication practices. Changing our communication style is one of the most difficult challenges we face, but there are excellent ways to do it, and the rewards are immeasurable. It helps tremendously with partnering; in fact it is mandatory. But the best news of all is that our new communication skills help us in our other day-to-day work and in all sorts of social settings, and also do absolute wonders in our personal and family lives.

INTERPERSONAL COMMUNICATION SKILLS ARE CRITICAL

The simplest of changes makes a world of difference. For example, before partnering you might ask, "Did you do such and such?" Sounds okay, right? If the person has done it, then they can answer as such, but you leave them with the feeling that you are checking up on them or challenging their performance. If the person has not done it, then they have to say "no" and they are "guilty." Their only hope is to try to explain their failure as quickly as possible and with as many reasons as possible. The phrase "Did you . . ." really puts them on the spot.

However, partnering communication suggests that you might say, "Did you get a chance to do such and such?" Does that really make a difference? Absolutely! If they did it, they can tell you they did and your only question was whether or not they had an opportunity to do it—i.e., you were asking if the world and all that is going on in it left them the time and opportunity they needed. If they had to do the impossible to get it done, you have invited them to tell you about it, and they will appreciate your asking. The crucial difference is that you aren't challenging them and their performance, you are simply asking whether or not they had an opportunity to do it. If the person has not done it, then they can say so and your question invites them to let you know what got in their way. Even if they should have gotten it done but didn't, the second phrase lets them save face while they let you know what's going on. "Did you . . ." sets you up as the challenger and the good guy, and they are either a bad guy or at best okay. "Did you get a chance to . . ." does not directly challenge them; it is a friendly inquiry into how things are going. Such simple changes will do wonders at work and miracles at home. These changes that help people get along better are a part of what partnering is about.

In the early stages of partnering, you and your group should do regular interpersonal communication assessments on yourself and each other and work through a number of open and lighthearted interpersonal communication workshops. When some individuals have difficulty picking up the new skills, they should solicit the help of all the others to help them catch and work on their slip-ups. If they say, "Did you . . . ," their other partners can simply grin and reply, "Do you mean, did I get a chance to . . . ?" You can have a good laugh among yourselves and get on with it. No big deal, but after a while the improved communication style catches on and sinks in with everyone. The partnering group is better off, and the individuals are better off. It is also a part of the group experience in helping each other grow that strengthens partnering.

DEALING WITH SENSITIVE AND CRITICAL ISSUES IS A MUST

The third most likely cause of partnering failure, and the last one we will talk about in this section, lies in skirting the most sensitive or difficult issues. It is okay to skirt an issue the first time around; that can be part of a good process that eases into an explosive situation rather than barging into it. But partnering is based on an ability and a willingness to be open, honest, and to behave in the best interests of the group and the project. Skirting issues is at best a personal compromise and is definitely not in the group's or the project's best interest.

Participants usually skirt issues because they don't know how to deal with them and they are convinced that they are better off skirting the issue than dealing with it. Having ways to raise and deal with issues is part of the partnering plan, and interpersonal communications training is key to effectively dealing with this challenge. If individuals in the partnering group learn that the difficult issues can be raised and dealt with to everyone's long-term advantage, they will do so. When the skills for raising and dealing with difficult issues are lacking, and the expectation is that the individuals will be worse off for having raised them, then the issues will be skirted and the fiber of partnering will begin to unravel.

At regular intervals in partnering you will be having roundtable sessions on what is working well, what is working okay but needs some work, and what isn't working as it should. One part of this session is to identify any difficult areas that may have been skirted and to gently get them back on the agenda. In the early phases of partnering it will be fairly common to have to face up to issues that have been skirted. But as each one of these early failures is nudged into the limelight and dealt with, the group will get more and more comfortable with raising and dealing with difficult issues. In time such skills become just another day-to-day part of good partnering and are handled as they come up rather than at the roundtable sessions.

A final word to the wise: It is an old ball game played a whole new way. Join in, but remember to come in with your head up and your eyes wide open.

4

What Is Partnering?

Partnering is not words, not charters, not contracts, not charts, not reports, and not meetings. It is a group charisma and an aura; you can feel it, there is partnering in the air. Partnering is to business as championship teaming is to sports. In sports there is so much more to the team than a field, rules, policies, procedures, forms, meetings, coaches, players, and plays. It is a special excitement and a special energy. It is being a part of something that outsiders really don't understand unless they have been there. When you tell outsiders about it, most people just say, "Uh huh, I see, sounds nice," and you know they really don't have a clue.

In another sense, partnering is a liberating process. It helps the best that can be to emerge from the weight of tradition. It is worth repeating that there is a greatness in people and companies and projects that is not fully tapped. It has been there for years, and for years it has been struggling to break free and do the extraordinary. But break free from what? How about breaking free from tradition, protocol, bureaucratic policies and procedures, litigation paranoia, and the detrimental consequences of stepping out of line from the masses or stumbling along the way. Partnering faces those constraints and aims to overcome them and set the best in us free to accomplish the extraordinary.

THE ROLE OF OPEN VERBAL COMMUNICATION

Essential and central to partnering is flat-out open communication: the truth, the whole truth, and nothing but the truth. However, the communication is of a friendly nature, not a mandated or adversarial one. Partnering is based on open communication and on wanting to share ideas, concerns, and information. The truth means no fibs, no white lies, just straightforward talk. The whole truth means no holding back just in case, no hidden agendas or withheld information, no hesitancy to avoid ridicule or a problem, just straightforward talk. And nothing but the truth means no exaggerations, no fluff, no meaningless gab to fill empty space or throw people off, just straightforward talk. Again, partnering relies heavily on the truth, the whole truth, and nothing but the truth, so help your project.

We look at the Japanese with their propensity for verbal arrangements between entities and their obvious blending of public and private interests in pursuing markets, and we can see that they do a form of national partnering, and on the whole it works pretty well. Granted, they do partnering to a level that goes beyond what our antitrust laws might allow, but there are probably some partnering lessons there for us.

ANCIENT PARTNERING TO MODERN PARTNERING

Partnering is certainly not a new phenomenon. When we were tribes roaming the earth, we were very good at partnering. We all chipped in and did the best we could for the clan. When something unusual came up, we did the best we could to respond as a group to survive. Partnering was an absolute necessity for survival. Groups that did anything less than be the best they could would fall under the evolutionary ax. The evolutionary ax had two blades: survival of the fittest, elimination of the unfit.

But there came a time when our survival didn't depend so much on being the absolute fittest. We could be pretty fit and still survive. We reached a point where total clan partnering for food and survival wasn't as important as it once was. We reached a point where we could build fiefdoms and bureaucracies and still get along. And here we are. But partnering still works, and it still puts its practitioners on top of the heap. With technological evolution emerging at a neck-breaking pace, with worldwide competition opening doors everywhere to the fastest and best who want to compete, partnering is coming back into vogue. In the business world we are rapidly moving into a new era of survival of the fittest, elimination of the unfit. Honoring tradition, protocols, and policy to the detriment of being the best we can be has ever sharper consequences. We have been in search of excellence, and it is coming on strong. The more massive the resources involved, the greater the time line; the bigger the rewards, the tougher the survival. The engineering and construction industry is facing one of the fastest reemergences of survival-of-the-fittest pressure. Getting back to basic partnering is an increasingly important key to long-term survival in our engineering and construction industry.

Some modern-day cultures and nations do partnering now (as with the Japanese), and in some lines of business partnering has been the backbone of their survival and growth. This is particularly true for many areas of small business that have banded together and helped each other beyond normal business practices to keep their enterprises afloat. In some ways it is a little curious that we in big business are finding that we need to return to acting like small businesses in order to survive. It is also curious to note that small businesses that used to work freely with other businesses grew and at some point outgrew the partnering style. And now we are discovering that the grown-up organizations cannot compete as well as they must if they are to survive.

The writing is on the wall. In order to compete, especially in the EPC environment, we must return to a partnering style; yet, as an industry, we have forgotten how. It is a strange turn of events, but a turn of events it is.

A RETURN TO TRUSTING IN PEOPLE

When our enterprises were small, unsophisticated, and engaged in partnering, we trusted in people. People got together and came to an agreement on what needed to happen. People exchanged ideas and concerns and worked things out for the best result. As we grew, we became sophisticated and began to trust concepts, plans, specifications, contracts, systems, inspections, and controls more than we trusted people. The systems had to be engineered and designed to mesh together to cause the end result to happen. As our systems and controls became more sophisticated and onerous, we as individuals pulled back and became self-centered. The systems called the shots. Join in, fall in line, or there was another system to take care of your removal. It was no longer a partnering "we" out to do the best we could.

The watch-out-for-me orientation combined with trusting systems and controls instead of people has created a heavy load of baggage that we carry around with us from project to project. In growing up and away from people-based business we have taken on a huge load of things that go beyond just what is needed to get the job done, things that get in our way, like excessive doses of policies, procedures, reports, inspections, protocols, egos, incompetencies, litigation paranoia, power plays, and traditions. When these things get in our way, we grumble and struggle, but we usually give in. It has gotten to the point where it is as if we as people don't build our projects. All those things that are a part of the system run us, and somehow it is the system that builds the projects through us. We just work for the system.

When we look at the final economics of our companies and projects, in truth most of these things are useful and pay their way (see Figure 4.1). We obviously need systems and controls, especially relating to protecting our national interests and the safety of our workers and customers. Policies and procedures that deal with deterring crime are essential as are those rules and regulations that ensure that we do not damage society as a whole. But at the same time that many of these things pay their way, many others do just the opposite. The problem is that even the detrimental processes are a part of a self-sustaining system. When we start to fight the things that work against us, they fight back. Even when reasonable people see what is wrong and what would work better, breaking and changing

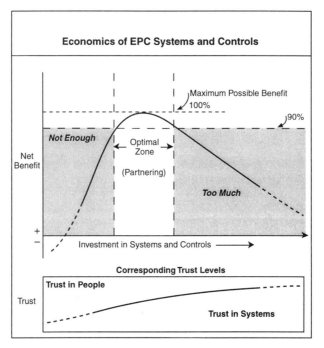

Figure 4.1.

the system is usually more than we can handle. It is usually better to just work with the system and do what we can. We know we could do better, but it seems that it is just not within our reach to do so.

PARTNERING IS A HURDLE BUSTER

As a nation, as companies, and as individuals, most of us sense that we could accomplish so much more if we could somehow move past, around, under, through, or over these hurdles. If we could get a fresh start on our working culture and style, then we could do so much more. TQM and reengineering are doing just that for work practices. Partnering offers a similar fresh start for us in the EPC business. Partnering is like TQM and reengineering for our project work-style. However, in partnering we don't exactly seek to stabilize a process, and we aren't looking to start from scratch. We are looking to wipe away outdated and meaningless hindrances in the way of extraordinary performance. Partnering is a management approach that aims to find the best path to the best performance and result. The goal is to do what it takes to open the doors to extraordinary accomplishment, not to fix any particular process.

Partnering is not usually out to fix the system. Its goal is to accomplish the most it can toward its objectives. When in its pursuit of extraordinary accomplishments partnering comes up against illogical, unfounded, or unreasonable hindrances, it first seeks to move by the hindrances as gracefully, ethically, and legally as possible. When confrontation is needed to deal with a hindrance, partnering participants are especially skilled, tactful, and determined to resolve the situation to best benefit their goal of extraordinary accomplishment. Constructive confrontation and resolving hindrances to accomplishment are at the forefront of successful partnering.

In their pursuit of extraordinary accomplishment, partnering groups often find themselves in the difficult position of needing to challenge well-entrenched and formal programs, such things as contractual do's and don'ts, management by objective goals, career advancement pressures, and individual or group merit and bonus programs. These generally beneficial forces and influences can also work against us and can stand in the way of creativity and extraordinary accomplishment. Using partnering to work through these problems and open new doors to new ways to get work done creates an exciting opportunity. It frequently happens that individuals become so energized by partnering and its momentum that they put their own career status in jeopardy for the sake of the project. Although we find some degree of such dedication in nearly all projects, we seem to find it more in partnering. There is a refreshing magic in partnering's new ways, and it quickly pulls in those who are most drawn to the camaraderie of extraordinary accomplishment.

Obviously partnering does not win all of its constructive confrontation battles. The individuals and the group do the best they can to rally the good forces and diminish the counter-forces, but there are limits. When the cost of gaining ground exceeds the benefit to be gained, the group refocuses on the ultimate objectives and moves forward. To do the most the quickest with the least, the group must remain realistic. Bogging down in personal or principle battles is not in line with partnering. Extraordinary accomplishment is not perfect accomplishment or ideal accomplishment; it is just extraordinary accomplishment.

PARTNERING CAN COME FROM ANYWHERE, CAN WORK ANYWHERE

In this book we are looking at partnering in the engineering, procurement, and construction (EPC) industry, especially major EPC projects, but the application and potential of partnering are much more universal. What we are talking about is a work style and value set that optimize the accomplishments of ordinary people in spite of traditional social and organizational counter pressures.

Partnering can emanate from almost anywhere and reach out in almost any direction. It can apply to customers, suppliers, field versus headquarters, interdepartment operations, auditing, accounting, systems, labor unions, and regulating agencies. Partnering will clearly be shaped by the contracts and legal entities involved, but it can thrive in almost any situation. As emphasized earlier, in any group, in any setting, there is a greatness in people and groups that is not being fully tapped. Partnering is an opportunity for ordinary and extraordinary people to accomplish the extraordinary anywhere, anytime. We are focusing on EPC, but we are talking about a universally valuable work style.

Partnering will and does work, in schools, factories, hospitals, agencies, communities, churches, shopping malls, grocery stores, and the military. The more you use partnering in as many different settings as possible, the better you will be for your efforts. And with your successes, the better others will be for joining you in your efforts. It is a win-win opportunity. Only outmoded practices, behaviors, and values are at risk, and no one really misses those when they are gone.

Partnering is teaming that goes above and beyond companies, agencies, and people just doing their jobs well. Partnering involves a special camaraderie and an unusual enthusiasm for doing what it takes to accomplish the extraordinary. At partnering's foundation is a charter among people by which they commit to behaviors that transcend simply good practices. The partnering members vow to be open, honest, and straightforward in sharing their ideas, concerns, and information. The members vow to vigorously seek new and better ways to support each other and achieve the most that is possible. The group agrees to persevere against the resistance it will face from those who value policies, procedures, traditions, and protocol more than they value changing for the better.

ESTABLISHING A PARTNERING GROUP

Partnering is a process that a group of people can adopt in order to achieve the extraordinary. For the rest of the chapter we will not distinguish between entity partnering, as in alliances, and personal partnering, as in a project team. As a business process, partnering has inputs, processes, resources, and outputs that must optimize performance within ethical, legal, and business bounds. Developing and managing partnering requires process engineering to design it, testing that can show how it is doing, and a long series of planning, doing, checking, and adapting cycles to ensure that implementation, delivery of value, and stability of process are properly achieved.

On the one hand, establishing partnering is very much like establishing or overlaying any other functional group into an organization. It is a business process challenge; you accept it, you plan it, you do it, you measure its progress and effectiveness, you keep it focused on success, and you deal with its shortcomings. On the other hand, it involves personal exposure and values that go beyond the normal call of duty. It is much more personal and thereby much more difficult than overlaying any other functional group into an organization. However, in this section we will start by discussing the business aspects of partnering and get into the tougher personal and interpersonal issues later.

When setting up partnering, the group is told that being creative and taking a few chances to achieve the extraordinary is more than just okay, it is expected. It is one of their primary challenges. It is an accepted norm for the group, so the charter says. But members of the group have heard that one before. They know that glory will no doubt greet the first few successes. But they also know that all hell usually breaks loose when there are a couple of setbacks. The hype in the beginning quickly steps aside for the reality of management's aversion to bad news and setbacks.

In the beginning it is easy to put into the partnering charter that, in order to make the most progress, the group may have to set a fast pace and take risky steps. It makes perfectly good sense that the group will plan to accept a couple of setbacks in order to gain a series of breakthrough successes. Easy to say. But when the second or third setbacks happen, will the partnering group be allowed to stay the course? Or are they forced back to a more conservative pace of safer steps with only an occasional uncomfortable setback?

Partnering requires an oversight process to ensure that its charter is alive and well. In this case, the oversight process must ensure that a faster pace is being taken, that the faster pace is yielding a better net gain, and that forces that are paranoid about a few setbacks are not derailing the partnering effort. With this first example we are making the point that partnering is a business process, and as such it needs a plan of expectations, a way to test results, and a regular cycle of adapting to circumstances and replanning new efforts. In many respects, running partnering is no different than running a service business. If you want it to succeed, you must design it and you must manage it. It won't happen just because you say it should. And it won't happen by happenstance. But it can happen with careful planning and execution.

Partnering's pursuit of best results applies not only to the end product but also to its individuals. Along with an oversight process for the business of partnering, the group also needs an oversight system to cover personal and interpersonal aspects of its members. This part of the oversight process regularly tests for how well individual participants are doing, where they may be falling down, and what needs to happen to get them back on track.

When it comes to open and forthright interpersonal communication, no hidden agendas, enthusiasm for the extraordinary, willingness to challenge the system, and ready acceptance of constructive criticism, some partnering members come along quickly, some come along slowly, and some never come along at all. Partnering's strength, potential, and value come from its membership. It must do its utmost to help those who fit in and set aside those who do not. A partnering team must do all that it can to leverage the capabilities of its members to the maximum and deal with those who fall short. Failing to assertively optimize on its personal and interpersonal power will seriously undermine the effectiveness of a partnering group and will generally lead to its downfall. Partnering does not tolerate less than the best in those it works with, and it does not tolerate less than the best in its own members.

DECIDING WHETHER OR NOT PARTNERING EXISTS IN A GROUP

It is time for a long-winded sentence you can use to decide whether or not partnering exists in a particular situation. For our purposes in this chapter, partnering potential exists when a group of business associates (1) has fully agreed on an

Scoring:
 	
110% Outstanding
100% Pretty good
90% Needs some
    improvement
80% Needs a lot of
    improvement
	Partnering potential is present when the parties to the contract attest that they are comfortable that they...

___ %	Have fully agreed on an ultimate mutual goal...
___ %	Have committed to a charter of essential philosophies, approaches and values for achieving the extraordinary, and...
___ %	Have launched a perpetual oversight process for regularly testing and improving their personal, interpersonal and group effectiveness against the goal and charter.
(1) x (2) x (3) = ☚ ___ %	**YOUR POTENTIAL FOR PARTNERING SUCCESS**

Figure 4.2. *Testing for partnering success potential.*

ultimate mutual goal; (2) has committed to a charter of essential philosophies, approaches, and values for achieving the extraordinary; and (3) has launched a perpetual oversight process for regularly testing and improving its personal, interpersonal, and group effectiveness against the goal and charter (see Figure 4.2).

The good news is that for these views we are not looking for perfection, we are only looking at where we stand on a sliding scale. Obviously there needs to be a reasonable level of experience and competence in those involved. You can use a percentage score for each where 80 percent means a lot of improvement is needed, 90 percent means some improvement is needed, 100 percent means pretty good, and 110 percent means outstanding. The catch is in knowing what really good partnering looks like in order to give yourself a valid score. That insight will come later, but for now a rough guess is good enough. The bad news is that your overall score is not an average of the individual scores. Each score is roughly indicative of your chances of success in accomplishing the extraordinary. If you are just starting, are optimistic, and you think you are at 80 percent on each count, your present potential for success is roughly $0.8 \times 0.8 \times 0.8$, or 51.2 percent. To make the bad news even worse, not many partnering groups hit a score that high. Without considerable experience and professional help, most groups are lucky to hit $0.5 \times 0.5 \times 0.5$, or 12.5 percent. That may be depressing on the surface, but here comes better news. With professional help and working the process aggressively, the path to high success levels is there and it is doable. You can climb the rungs on each score and with time and much effort get to the higher levels. That can mean big rewards for you and your company.

Let us work through a couple of examples. If extraordinary performance in your situation could mean your project could come in 90 days early and $60 million under budget, then if you are at the 12.5 percent level you are looking at 11 days early and $7.5 million under budget. That's not bad. But read this book, work your tail off, get up to the 51.2 percent level, and you could be looking at 46 days early and $30.7 million under budget. Not a bad return for your effort! Although we have used only time and installed cost in this example, partnering typically makes equally impressive breakthroughs in safety, environmental, and life-cycle economic measures.

PARTNERING DOCUMENTATION

So far we have been setting the stage for partnering. Now we turn our discussion to what partnering is by noting what documents are used in partnering.

Partnering Charter

First and foremost is usually some form of charter or agreement. It is usually not so much a legally binding document as it is an informal agreement between friendly forces of one or more entities that are charged with working together to their mutual benefit. It is a little like the first couple of pages in a scouting book or the attestations of a religious order. It describes a common goal that all swear to support and sets down in writing the ground rules that should lead to mutual accomplishment of the goal and a satisfying experience for all the members. When we join we intend to comply fully with the charter, but if we fall short we aren't likely to get sued. And in the end performance and success are matters of pride and enthusiasm for what is good for ourselves and the group, and not so much a matter of complying with laws or edicts.

The basic components of a partnering charter involve defining a common mission, stating common philosophies and values, and outlining behaviors and practices that are central to the partnering effort.

A charter is critical as a touchstone for partnering, especially as partnering is usually a nonstandard practice operating as an overlay to standard business agreements and practices. However, it is a big mistake to simply adopt another group's partnering charter and move forward from there. Deciding what will and will not be in the charter is an essential aspect of building the partnering team. Struggling through what wording is and is not acceptable to each other is an even more important part of building the partnering team. The charter is not only a touchstone for the partnering group, it is also a certificate of accomplishment for the first phase of partnering. It is the group's first merit badge, and each group must earn its own as a foundation for what is ahead. Using someone else's merit badge won't do, and neither will using someone else's charter. We will talk more about earning this first merit badge when we talk about the partnering process in a later chapter.

Partnering Handbook

There is a lot to understand about partnering, especially how it works, what makes it work well, what undermines its performance, how to do partnering on a day-to-day basis, how to develop good partnering judgment, and how to handle the most common opportunities or challenges that are sure to come up. Each group will be well served by compiling a compendium of training materials, its charter, synopses of meetings, reference articles, and the like.

Group Assessment

Key aspects of the charter need to be converted into a survey (test) format so that from time to time the group can assess how well it is or is not following the charter. The partnering assessment form is essential to affirming what is going well as a way to perpetuate and increase success.

High positive scores provide defensible evidence that celebrating is deserved and in order. The instrument must also identify what is not going well and what requires the group's attention. More than with most surveys, the partnering survey bluntly looks for the negatives that might be hidden. The benefits of things going right are substantial, the consequences of things going poorly are just as substantial.

When the instrument looks for the negative and finds it, the required attention is obvious. When the instrument looks for the positive, if the score is low the group can look into the matter. As the partnering philosophy suggests, the group does not shy away from looking straight at problems and shortfalls and goes right to the heart of whether or not there is a problem.

The same instrument, or one only slightly reworded, is also vital as a feedback mechanism for how the partnering efforts are being viewed by those the group works with. When differences in perspectives between the partnering group and those it works with or affects cannot reasonably be explained, the group will need to determine if it is out of sync with the real world. The partnering group may be seen as reckless or pushy by those who are not involved and don't understand; this can be part of a natural reaction. But if the group is seen as detrimental to key aspects of the project (safety, schedule, quality, life-cycle cost), then a serious investigation by a professional is certainly in order. Those outside the group may hold a poor opinion of the partnering group and its methods, but they should almost all agree that the project is benefiting from what the partnering group is accomplishing.

Member Assessment

A form that is quite similar to the partnering assessment instrument should be developed to specifically test what each individual member thinks of their own participation and the participation of each of the other members. These forms obviously probe very personal aspects of the members. Even the most mature partnering groups should not attempt to collect, assess, and provide feedback on this type of material without the help of professionals. In the first place, time spent collecting and assessing the material and preparing for and facilitating feedback sessions detracts from the team's real work and value. The partnering members' efforts are best invested in doing what they should be doing toward extraordinary accomplishment. The professional facilitators can handle the partnering assessment at the same time they handle the partnering member assessment, so the two should be combined in one comprehensive package.

DIFFERENT VIEWS OF PARTNERING

Partnering is based in reality. Reality is at the core of the truth, the whole truth, and nothing but the truth. In the pursuit of truth, individuals in a partnering group become good detectives. The group learns to quickly sift through symptoms to find root causes. Their pleasure and success lies in correcting root causes, not just patching over symptoms. They also learn to quickly sift through the double-talk that people throw at them. When they hear "We can't do that," "It's against policy," "I'll have to check with my boss," and such, they keep asking "Why" until they get to the real reason or they at least blow past the puffery to the real issue. Not sifting through symptoms to root causes and not sifting through puffery to the truth are key reasons why so much baggage is still with us. When baggage is in partnering's way, partnering wades through it and does what it can to clean house. Cleaning house and setting things straight are important aspects of what partnering is about.

On the surface, partnering may appear to be an easier, kinder, gentler working style. And in many ways it is because of its open communication and efforts to eliminate complications caused by practices and behaviors that are at cross-purposes to the goals at hand. But in that same sense, partnering is tougher than most work styles. Falling back on excuses is the easy way, not challenging the system is the easy way, not probing for the truth is the easy way, looking the other way when things get messed up is the easy way, hiding things with the hope that they may not be discovered is the easy way, not challenging your friends is the easy way, and not confronting counterproductive behaviors is the easy way. Partnering accepts none of these paths, so it is not easy. Partnering is not timid about raising difficult, delicate, or sensitive issues, so it is not especially kind. And partnering doggedly and vigorously pressures whatever stands in the way of extraordinary performance, so it is not gentle by any measure. For those who take easily to partnering, it may be an easier, kinder, gentler working style. But for those to whom partnering comes with more difficulty, or when individuals get in the way of partnering, they will find that partnering is tougher and harsher.

The tougher and harsher edge of partnering is toned down by a higher degree of interpersonal skills practiced by the partnering participants. With training "You're wrong" becomes "I guess we see this differently, what I thought was going to happen was . . ." "That's stupid" becomes "If we do it that way I believe it will have a significant time and cost impact on the project, is there any way you can help me . . . ?" Another favorite tool of the partnering group is an old standby: Focus on the behavior or event, not the person. The goal of partnering is to achieve extraordinary results, not correct or defeat every one who resists along the way. Destructive confrontation absorbs energy, builds higher barriers, saps creativity, and creates whole new causes and battles that detract from the mission. Partnering is loyal to the end product first and foremost, and constructive confrontation best serves that end. Constructive confrontation requires an unusual breadth and depth of interpersonal skills, thus partnering requires an unusual breadth and depth of interpersonal skills.

From time to time you will hear partnering compared to marriage. Although marriage may be a reasonable comparison and model, the comparison doesn't sit well with a lot of us. It is true that to be successful we have to be honest with each other, we have to freely offer help and willingly accept it, and we have to get our issues out on the table where we can deal with them. It is also true that from time to time we may yell at each other and then must follow a commitment to resolve the situation. But for most of us a marriage model has too much personal and intimate baggage attached to it. For most of us partnering is a much different circumstance than two people working things out in a marriage.

PARTNERING AS A TEAM SPORT

Comparing partnering to a national team sport seems to make more sense. In national sports there are owners, coaches, special teams, and highly visible and well-paid stars. There are practically invisible workhorses that shore up the team, equipment and facility managers, hoards of suppliers and sponsors, and paying customers that need to be satisfied or the sport industry falls apart. In the sport model the players also have to be qualified, willing to learn, willing to give it their all, willing to put their personal interests second to the team's, and if they are going to make it to the finals they must share their best ideas, concerns, and thoughts to help the team be the best it can be. Getting serious about partnering is very much like getting serious about having a winning ball club.

In national team sports, if you don't have good players, good contracts, and good teaming, you will never get to the finals. You will have so-so pay, so-so crowds, so-so satisfaction, so-so accomplishments, and you will forever be dreaming of being champions but never really get there. However, having good players, good contracts, and good teaming won't guarantee getting you to the finals, and it certainly won't get you to the top of the heap. To be a championship team requires extraordinary teamwork, commitment, hard work, enthusiasm, vision, and honesty in putting your aspirations and shortcomings right out in the light of day. Being a champion team requires something extraordinary. Partnering requires no less.

Some argue that partnering is not that big a deal. Good people everywhere dig in and get the job done. The doubters point out that good people are the key, not partnering. To a certain extent they are right, but on an important and critical point they are coming up just a little short. As with almost all cases of extraordinary accomplishment, a lot of people can do okay, and some can even do quite well. But only those who squarely face every hurdle and expertly work through every detail can get to the big prize. It is that extra little bit of vigor, that extra little bit of perseverance that yields the last ounce of performance. And it is that last ounce of skill and performance that is needed to put them on top and in the winner's circle. In every race and every sport, the big prizes and honors go to the winners. Second and third place do okay, but it's nothing like the winner's status. And yet, what is the difference in performance between first, second, and third? After a two-minute race the glory goes to the one that is one-or two-tenths of a second ahead of the others. After battling it out nip and tuck for an hour, it is often just a matter of a couple of plays out of a hundred or more that determines who gets the prizes and endorsements.

In business, after spending millions of dollars, the difference between medium and high returns may rest on just a couple of percentage points in costs. Getting major performance bonuses on a mega project may be a matter of just a few days out of a three-year time frame. Getting a project up and running on schedule instead of three months late may boil down to just three or four open and creative partnering meetings at the beginning of a project. In every business it is that last little bit that makes a big difference in where you come out, and in engineering and construction partnering can be that last little bit of difference. Do without it and you may do well, but you will probably be looking at the winner's circle from the outside in, not the inside out.

MEASURING EXTRAORDINARY ACCOMPLISHMENT AGAINST THE MISSION

Partnering is more than just working as one. Partnering is working as one toward a single mission. On the surface having a single vision may sound simple enough, but in practice it proves to be one of the most difficult aspects of setting up and managing a partnering group. In order for the partnering participants to have a single-mindedness about extraordinary accomplishment, they must have a way of determining exactly which options to follow to maximize their efforts. But what is involved in selecting options? Is it cost? Is it efficiency? Is it performance to specifications? Is it life-cycle benefit-to-cost? Is it timeliness? Is it safety? Is it net present value, return on assets, return on equity, or earnings per share?

When partnering groups need to choose between their options or settle their differences, how do they measure which is best? When they need to convince others to change their practices, policies, or protocols, how does the partnering group convincingly justify the new over the existing? What is the equation that drives and focuses partnering? What measure best serves to guide extraordinary effort to extraordinary accomplishment?

In truth this is an area of partnering that is still struggling. In most cases partnering has to use the same tried-and-true performance measures that we've always used: cost, quality, safety, and timeliness. Partnering generally starts with a fair amount of time spent on studying and developing a mission statement that makes sense to the team. Right up front the group identifies where goals, objectives, specifications, and requirements seem to be at odds with each other. Some contention between goals is a given, such as holding down costs while promoting safety, or minimizing total installed costs while optimizing life-cycle economics. In most cases the partnering group works very hard at understanding the mission statement, and then appoints itself as judge and jury for sorting out options and making choices. The judge-and-jury role clearly cannot be taken lightly.

Early in their forming process partnering groups spend significant time understanding how their work and mission statement will affect themselves, their organizations, their companies, those who have already worked on the project, those who are working on the project, those who have yet to work on the project, those who will take ownership of the final work, and those who will support, work in, or be served by the final project. Just as a judge and jury are charged with representing the laws and principles of the land and upholding the reasonable-person concept, the partnering group is charged with representing all the others who must be considered in making project choices. The sharp focus on a single measure that can lead to extraordinary accomplishment is not after all an equation; it is a decision system. It is a continually honed judgment system within partnering that in many ways transcends what we are used to using in our day-to-day workplace. Because of its lack of precision, such a nebulous system would seem to be ineffective. Yet, because partnering recognizes it as the nebulous system that it is and pursues it with great seriousness and care, it is much more effective than the usual set of project measures. Being the caretakers of such a weighty endeavor causes the group to be especially careful and creates a tremendous feeling of responsibility and importance within the group. The partnering participants are charged with making the best decisions possible for a wide range of interests. It is a very stimulating responsibility that deepens the participants' commitment to performing to their best.

BREAKTHROUGH THINKING AND INTESTINAL FORTITUDE

Partnering involves relentless "possibility" thinking. In every meeting, in every encounter, and in every participant's behaviors, you can sense essentially the same question: What is the best possible way to get the most possible done in the quickest and least costly way? Developing a knack and a habit for looking for the best possible approaches is at the foundation of successful partnering.

Reality certainly doesn't allow partnering to always get its way, but over time the group gets more and more creative in devising better and better ways, and it gets more and more effective at getting its way. Champion sports teams and world class orchestras don't get there overnight, and neither does partnering. Even the very best highly driven talent still needs time and practice to fine-tune its performance to be first-rate. In that regard partnering is no different. The key lies in understanding what it takes to become a champion team and dedicating what it takes to get there. No doubt there have been numerous champion-grade teams that never made it because they didn't stay the course. They focused more on their stumbling than their goal, and they ended up not reaching their goal. If you have the capability and stay the course, then you can get there. Almost every partnering team has the capability, but few really believe in the possibilities and stay the course through those first tough weeks.

Breakthroughs from partnering do not come easily. As a rough rule of thumb, the group will experience about seven setbacks for every breakthrough they score. The bottom line is that the breakthroughs can save big bucks and time. One breakthrough easily more than makes up the time, effort, and expense involved in the seven setbacks, but it takes a lot to hang in there through the seven setbacks to get to the breakthrough. It takes a special mindset and determination that have to be developed and fiercely maintained. Not everyone can handle that sort of responsibility. It is somewhat like the venture capital business. Even if they have the money to waste, most people can't stand losing several million dollars on several ventures in a row. However, the venture capitalist has faith that every so often they will score big. One in six, or one in eight, or one in ten will yield a 20-to-1 or a 40-to-1 return, and the venture capitalist will be smiling all the way to the bank. The setbacks along the way pale next to the successes. Partnering is a little like running a venture capital business but works through a portfolio of ideas rather than ventures. Each idea is tested, given a shot, and after taking a few lumps here and there the breakthroughs come along and the setbacks are quickly forgotten. But partnering can't have the breakthroughs without the setbacks; thus a stomach for the setbacks is an important part of the process.

Perhaps one of the most important lessons we can share with our readers is that partnering success does not belong to the most talented or the best prepared or the best organized, it belongs to those who see it through. Staying the course and seeing it through will resolve talent, preparation, and organization issues. When it reaches a critical momentum, partnering is self-directing, self-diagnosing, self-correcting, and self-motivating, but only if it stays alive long enough to hit that critical momentum.

Stop pushing just a few days short of the critical point and the failure will be just as big as if it had never started at all. Hit that critical point and you can hardly shut it off. Once you have been through the birth of a partnering group, you not only pick up the faith that your next experience will make it, but you learn to tell when you are getting close to the critical momentum point. However, until you have been through your first partnering success, you can't really tell if you are a few days or a few weeks away from the critical point. Working with experienced people and facilitators will greatly help you understand and get on the winning side of the critical-momentum issue.

5

How Does Partnering Apply to Owners?

Partnering is essentially a working philosophy and a management style. Therefore, partnering is not dependent on what forms of entities are involved or which entities are involved. Whoever chooses to participate can accomplish the extraordinary under their own circumstances whatever they are, and that subset of partnering can be a success. However, the overall magnitude of success for a partnering effort does depend heavily on who participates and who does not. In EPC work the owner's interests are central and absolutely critical to defining what will constitute the highest level of extraordinary accomplishment. Thus, whether the owners participate or not, they are an integral component of the success equation.

Partnering can and does produce extraordinary results when it works well, and it works best when all the key players sign on to do whatever makes the best sense to maximize the end result. That the owner is open and forthright in defining what constitutes extraordinary accomplishment is vital. That the owner cooperates in every way possible to achieve the extraordinary is also vital.

Partnering applies to owners to the extent that it can best serve their needs. By joining wholeheartedly in the partnering effort the owners are leveraging partnering to best serve themselves, and they are the key to leveraging partnering for the highest possible level of extraordinary accomplishment. In short, partnering applies to owners because they are a (if not the) prime beneficiary of the effort.

WHEN THE OWNER DOES NOT PARTICIPATE

Before we discuss in detail the owner's role in partnering, let us take a moment to consider what happens when the owner chooses not to participate. What then?

44

If the engineering and construction group form and pursue partnering without the owner, they can still formulate what will constitute extraordinary accomplishment, and they can still optimize their resources for the maximum progress at the least cost in the quickest way. They can still work through and around politics, policies, tradition, and protocol to get the most done in the best way. The most significant difference without the owner's participation is that the partnering group's focus and commitment is on its definition of extraordinary accomplishment sans the owner.

If the group can reduce total installed cost and be rewarded for doing so, it's a go. Rewards can come in many forms. The reward can be in the form of bonuses or future work from the owner, or the reward can be a favorable reputation within the industry. But in some way the group needs to be better off for having pulled off significant reductions in total installed cost. If the only significant result of lower total installed cost is less work, that hardly constitutes extraordinary accomplishment for the group, and its creativity and efforts will likely be directed in other directions.

The issues are the same if the group can complete the project sooner or improve life-cycle economics. The key is, what's in it for the group, or why do it? If the owner lays a contract on the table and assumes the old adversarial relationship by emphasizing compliance with the contract with no exceptions, that is what they will get. The partnering group's only choice in that situation is to do its magic to best reward itself. The total focus for extraordinary accomplishment will be to get the most mileage out of the project and move on. The owner will likely get little benefit, and the participants will gain only what circumstances allow.

WHEN THE OWNER DOES PARTICIPATE

With the owner's participation the opportunities for breakthroughs and extraordinary accomplishment are multiples of what they would be without the owner's participation. With reasonable sharing of the risks and benefits by everyone involved, the partnering group and owner can be fully aligned and two critical things can happen. The range of breakthrough possibilities widens dramatically, and the likelihood of success increases.

MEASURING EXTRAORDINARY ACCOMPLISHMENT POTENTIAL— A MODEL

Here is a simple model to help the reader understand what is at stake here. Let's say that we are looking at a $500 million EPC contract covering a two-year period. Our ordinary accomplishment would be to meet and satisfy the contract. Let's say that we have gazed into our crystal ball and believe that, with extraordinary accomplishment, partnering could complete the project three months early and at 95 percent of planned budget. The installed cost savings could be up to 5

Ideal = 100%	Areas to be scored:
____ %	Partnering effectiveness
____ %	Owner involvement
____ %	Equitable sharing of risks and rewards
(1) x (2) x (3) = ✍ ____ %	YOUR POTENTIAL FOR EXTRAORDINARY PARTNERING BENEFIT

Figure 5.1. *Estimating potential for partnering's accomplishment of the extraordinary.*

percent of $500 million ($25 million), and we will add that at a $125,000/day positive cash flow from the new facility, completing the project 90 days faster could be worth up to $11 million. For this crude model we are estimating that extraordinary accomplishment could therefore be worth roughly $36 million.

For this discussion, achieving the maximum benefit from extraordinary accomplishment requires three dimensions. Each dimension is scored on an effectiveness scale of 0 to 100 percent. The three dimensions are partnering effectiveness, owner involvement, and equitable sharing of risks and rewards (see Figure 5.1). However, the expected result is not a weighted average of these scores, as we are used to seeing. In this case our expected result is much more a result of the interdependencies among the dimensions, and so we use the product of the three scores.

For example, if we score 100 percent on all three dimensions we can expect the full $36 million benefit. If we score 80 percent on all three we are down to a product of 51.2 percent, or roughly $18.4 million in benefit. And if we score only 60 percent on partnering and owner involvement and only 25 percent on sharing of risks and rewards, we are down to a likely benefit of only just over $3 million (9 percent). The interdependencies among the three dimensions create a much more dynamic situation than when the scores might be essentially independent of each other and a weighted average might be appropriate.

GIVE AND TAKE OF OWNER PARTICIPATION

From this crude model we can see that if the owner's participation is fairly limited (say 20 percent, and if few provisions are made for sharing risks and rewards (say 20 percent), then the benefit from a 100 percent partnering effort is severely limited (4 percent). Under these circumstances it will be very difficult to come up with breakthrough ideas, to get people very excited about them, to implement them, and to stay the course to really accomplish much out of the ordinary against total installed cost and schedule.

However, the 100 percent partnering may work wonders for the EPC team, sans owner. The workers can gain in safety and satisfaction, and the contractors can do a nice job of protecting themselves and focusing on improving their individual situations (while meeting the requirements of the contract). The benefits to the owner may be limited to 4 percent ($1.4 million), but the side benefits to the other players may be relatively significant.

Owners who choose not to participate in partnering, or at least choose not to participate well, will certainly be limiting the benefits that they might otherwise enjoy. To the extent that they limit the big breakthroughs for the project, they are also limiting the benefits that the EPC contractor might have otherwise gained, and by not participating they are hurting their own situation far more than they are hurting their contractors who go ahead with partnering without them.

ISSUES BEHIND OWNERS NOT PARTICIPATING

Partnering applies to the EPC owners because they have the most to gain from it. It applies to the EPC owners because their participation is the major factor in how much benefit can be realized and to whom those benefits will flow. Partnering applies to the owners because not backing it is self-defeating. Making a case for owners to join in with partnering is much easier than making the case that they should not. And yet many owners still choose not to join in, and we need to understand why they make that choice.

Owners often choose not to partner because what they are doing appears to be working okay and designing and building projects is not their major focus. In today's markets keeping one step ahead of the competition is the lifeblood of staying alive. Getting improved products and services to the market faster and better than anyone else is where the really big pots of gold are. Keeping a tight rein on markets and margins is critical for keeping investments and resources aimed at the right targets. And maximizing capital structures at the right risk/ reward levels can make the difference between so-so endeavors and megabuck endeavors. So there is plenty for top executives to do besides get involved in the latest management fad.

The return on partnering in most cases pales in comparison to the return on most of these other endeavors. If all the executives are already maxed out on other more important things, then fair enough. Considered alone, partnering's return on investment usually isn't high enough to compete with the other core business big hitters. But can that really be the case? The executives who are responsible for the EPC projects are already responsible for the best possible outcome. So they should be interested because a more effective design and construction project will add to the returns anticipated by the business.

Some owners would partner in a minute if they really thought it would work. They aren't stupid; they wouldn't pass up another few million dollars in benefit if they knew it was there. For these people the real issue is that they do not believe they will be better off by partnering. They have tried lots of new things before.

Many new things are a disappointment; some are downright disasters. To avoid disappointments and personal disasters, these people simply have to be absolutely convinced before they will step from the tried and true to the new. For these hesitant owners we suggest personal interviews with those who claim partnering works, and especially careful planning and facilitation for their first effort.

Some owners simply don't understand the concept and clearly don't understand how they would personally be able to manage such an open-style, free-wheeling group. After all, most owners are not continually building new projects. Partnering might do some good, like some folks say, and it might rip your leg off, like some other folks say. In some cases some owners are too limited in their ability to "handle" such a group, and they will do what they must to keep from having to face or deal with the issue. In these cases owners can also be convinced through research and interviews that partnering is a smart approach for their situation.

In reality, owners don't always have to jump in with both feet. They can be up front about their concerns, discuss a limited role in the partnering process, and ease the other foot in when it makes sense to do so. Partnering doesn't have to be an all-or-nothing step for the owner. They can start with a 20 percent commitment and see how it goes. They can increase their involvement one step at a time when it feels right to do so and stop or back out when that seems to be the best course. In this regard, partnering applies to the owners only to the extent that it makes sense to them. However, it makes good sense that they should give it careful consideration and rule it out or restrain it only after a well-informed and carefully thought-out decision. As discussed elsewhere, there is hard evidence that partnering in most instances works impressively well. If EPC performance is important to any owner, then partnering certainly deserves at least a fair hearing.

SUBTLE BENEFITS TO OWNERS FROM PARTICIPATING

Partnering applies to owners in more ways than just installed cost, schedule, and life cycle economics. Along with delighting the owner in these ways, partnering also forces owners to sort through difficult issues and decisions that otherwise might be glossed over. Partnering's single-mindedness toward extraordinary accomplishment causes it to close in directly on turf issues, diversity issues, organizational weaknesses, unclear policies, and anything else that interferes with extraordinary accomplishment. Facing and resolving these issues to benefit the project at hand is sometimes a small part of the long-term value. Resolving these issues through partnering on one project may eventually improve operations in many areas, smooth the way for other improvements, and clear up issues before they become serious legal problems.

Significant companywide advancements and breakthroughs often get their start in small isolated programs like partnering. Clients who may believe that partnering had no marginal impact on their project can learn later that the ripple effect from the project did bring value to the company.

STAFFING AND TURNOVER ISSUES FOR OWNERS

The constant pursuit of extraordinary accomplishment is unbelievably invigorating. It is also exhausting. It seems, however, that most of us find the exhaustion from extraordinary accomplishment to be much easier to bear than the exhaustion from less-than-ordinary accomplishment or failure. The stress levels in partnering are high, but it is generally a good healthy stress as with teams that are in the final playoffs. Partnering participants also worry a lot, but they worry about missing a great creative opportunity or not being able to unravel a seemingly impossible situation for a really great breakthrough. The stress is there, but it is good stress. The long hours are there, but they lead to extraordinary accomplishment. The fuel for partnering is camaraderie and zeal, not defense and fear as is the case in so many situations.

It is clear that partnering is a very personal adventure. It involves mature interpersonal skills and a considerable amount of trust. Owners must commit to hold participation turnover to a minimum. The investment and the value are in specific people involved in specific working relationships. In most business situations it is a matter of good people in positions getting the job done. But in partnering it is much more than that. Because partnering's power comes from well-developed working relationships that transcend ordinary business, plugging people in and out is especially detrimental.

Some turnover is normal and acceptable. Too much turnover and the group looses momentum, creativity, and effectiveness. Partnering group turnover should be held to no more than say 20 percent in any one quarter and no more than perhaps 30 percent in a year. Any more turnover than that and maintaining partnering becomes too burdensome. Certainly, those rates are for discussion purposes only. Some groups are so strong that as long as the strong participants remain they could take a 40 percent or 50 percent turnover and survive and still produce good results. Other groups are so fragile that any turnover at all will cause a collapse. The issue is not in setting a turnover maximum, but in understanding that the group must be very sensitive to nonessential turnover and must stand ready to properly adapt when turnover happens. If it means holding a reforming session, so be it. The power is in the partnering group's dynamics. Those dynamics must be kept tuned for top performance. A partnering group that drifts out of tune will not lose its power at a linear rate; it will lose it at an exponential rate.

If turnover can be held to 10 percent or less per quarter, it is usually not a serious factor. However, as it climbs over 10 percent it becomes exponentially more difficult to absorb without material impact on the partnering effort. Except when housecleaning is necessary, the more stable the group the better. Here again partnering is much like a championship team; it can stand just so much turnover before it loses its magic and has to rebuild its strength.

DEALING WITH LOW BIDDERS

The industry must always face the problem of low bidders who may not be able to perform in the owner's best interests. During bidding, as long as the bidders appear to qualify, it is difficult to step away from the low bid, even if they are somewhat suspect. Choosing a higher bid begs an explanation to the stockholders, owners, or government interveners. In some cases partnering contractors will hopefully be the low bidder, or at least reasonably low, so that owners can choose the partnering contractor if they wish. However, the owner has a problem when a good partnering firm is not the low bidder, but the owner wants them to do the work. How can they justify not taking the low bid? Some owners can make a life-cycle economics argument based on their assumptions about how they think the EPC contractors will perform and can pick the higher-cost bid because they can make a case for equal or greater long-term benefits. But such an argument is difficult and will not fly in a great many situations.

The more practical way of dealing with this problem is to make partnering a bid requirement and be able to include and exclude contractors based on their partnering experience and credentials. This way low bidders without proven partnering performance can be eliminated. Low bidders with good partnering credentials may still enter the picture, but along with a partnering approach will likely come reward and penalty clauses that should protect the owner. And with partnering the owner will hopefully be able to spot and rectify a bad situation before it gets too serious.

CONSIDERATIONS WHEN HIRING A PARTNERING CONTRACTOR

Selecting an appropriate EPC contractor for partnering is a serious matter for the owner from another perspective. Really good EPC partnering means that the EPC contractor not only represents the owner's best interests in all things, but usually accepts responsibility for many roles that have traditionally been filled by the owner. An owner who gets deep into partnering may lose the ability to manage non-partnering work. Their resources that used to take care of enforcement, auditing, compliance testing, inspecting, and the like may fade away under extensive partnering. That is good and that is bad. Owners are usually better off not carrying contractor oversight staffs and systems if they can avoid it. If the EPC partnering contractor provides those services, then they are there and paid for only when they are needed, and they aren't surplus to the owner between contracts or during contracting lulls. If the EPC partnering contractors fulfill the traditional owner's roles, then the owner must take selecting their contractors very seriously, and they must make sure that their representatives on the partnering team are also ensuring that the EPC contractor is providing quality owner-substitute services. Does this mean the owner should return to big-time inspections again? Certainly not, at least not to the previous level.

Our point is that the owners must not abdicate their responsibility to the EPC contractor, even a well-known EPC partnering contractor. There will be a few EPC partnering contractors who for one reason or another will not deliver. The owner must be sufficiently involved to detect those situations and act on them. If partnering is working well, such EPC contractor failures should be few and relatively minor. But life is life, reality is reality, people are people, and problems happen, even under the best of circumstances. Owners can and should enjoy the benefits of partnering, but owners must also continue to be aware, alert, and engaged in their EPC work.

6

How Does Partnering Apply to the Engineering Construction Business?

In the late 1970s and early 1980s there were significant changes taking place in the United States that were to have a direct impact on the design construct industry.

- In the late 1970s large new nuclear power plant contracts declined for a multitude of reasons. This caused the beginning of the reduction in the related engineering and construction power organizations that were large, and sophisticated; the nuclear power business represented a significant portion of some corporations' workload and backlog.

- The book *In Search Of Excellence* was published, explaining how some firms had a competitive edge because of the way they operated. Peters and Waterman raised many of the soft issues of management in this writing.

- Many owners of engineering and construction projects, in large part as a result of deregulation, were becoming cost-conscious and looking more aggressively for ways to reduce their projects' total installed costs in order to be more competitive.

- In the early 1980s, the price of oil rose, making petroleum projects more economically feasible. The engineering and construction business staffed up in anticipation of additional work. However, the oil price decline in 1982/1983 led to cancellation of several of these large projects, and a downsizing in the engineering and construction industry followed.

- The general economic recession that started in 1982/1983 had a direct effect on delaying many engineering and construction projects.

- Internally, many companies outside of the engineering construction business that had built up engineering staffs began to look for ways to "outsource" the function.

- In 1984, Shell Oil entered into the first Construction Industry Institute–recognized "true" partnering agreement with SIP Engineering.
- In the early to mid-1980s, Deming's management methods began to gain recognition in the United States after his course book *Quality, Productivity, and Competitive Position* was published in 1982 and *The Deming Management Method* by Mary Walton was published in 1986. Innovation, continuous improvement, and quality became principles adopted by many companies.
- In 1987, the Construction Industry Institute established a task force on partnering to evaluate the feasibility of this method of doing business in the construction industry. There were seven major partnering alliances in place and working.
- In the late 1980s, the concept of total quality management (TQM) was developed and began to be implemented. (See Figure 6.1.)
- In 1991, the Construction Industry Institute published *In Search of Partnering Excellence,* which was prepared by a special partnering task force. At the time there existed over eighteen true partnering alliances. The study summarized its findings by saying:

Partnering is an excellent vehicle for attainment of TQM in the construction process. An effective partnering relationship will facilitate improved quality by replacing the adversarial atmosphere of a traditional business relationship with a

---

WHAT IS TOTAL QUALITY MANAGEMENT?

Total quality management (TQM) is a relatively new approach to the art of management. It seeks to improve product and service quality and increase customer satisfaction by restructuring traditional management practices. The application of TQM is unique to each organization that adopts such an approach.

—Report of the U.S. General Accounting Office, May 1991

It seems that what you're telling us boils down to five basic principles:

1. Ask customers what they want.
2. Set zero defects as your standard.
3. Complete work in the shortest possible time.
4. Measure the system, not personal behavior.
5. Make sure everyone feels like a stakeholder.

—U.S. Congressman Newt Gingrich
(*Making Quality Work,* George Labovitz, page 1.)

Figure 6.1

team approach to achieve common goals. Team members can challenge directives when the impact on the work affects quality or is disproportionate to the benefits. The potential for improved quality also is increased due to a better understanding of project scope and an atmosphere more conducive to implement new technologies.

- In 1988, the Army Corps of Engineers began using a variation of partnering that required team building, collaborative problem solving, continuous process improvement, and dispute avoidance for each contract awarded. The partnering concept used was so successful that in 1992 the Chief of Engineers promulgated a policy statement that partnering would be practiced on every new Corps construction contract as well as in other business relationships.

THE NATURE OF THE ENGINEERING AND CONSTRUCTION BUSINESS

The basic business of the engineering and construction firm is one providing some or all design, engineering, finance, procurement, construction, and project management services to accomplish projects with labor, equipment, materials, and sometimes other services.

The size of projects can range from a $20 billion industrial complex mega project to small projects ranging from $100,000 to $5 or $10 million. In addition, some engineering and construction firms have pursued operations and maintenance-type work when the individual work orders are small or very small, from $100,000 to $2 million, and are usually revamping or modernization work.

The design construct business is designing and constructing projects that have a discrete beginning and a discrete end and are usually unique. Engineering and construction projects are not continuous in the sense of a process or a manufacturing line, although there may be more opportunities to copy the lessons learned in manufacturing as design replication expands.

There are nine phases of an engineering and construction project:

- Feasibility
- Development
- Finance
- Concept development and review (preliminary engineering)
- Estimate
- Detailed engineering
- Procurement
- Construction
- Start-up

There are people involved in each phase of the project. Projects can be characterized as putting a puzzle together in four dimensions—length, width, depth, and time. Most engineering and construction projects by their nature are unique, and therefore every piece of the puzzle is a unique and one-time event. Each unique puzzle piece is like a problem to be solved. When one is solved, there is another one to solve until the project is completed. This differs from a manufacturing line where the same activity is ongoing and basically repetitious. In engineering and construction the processes to some extent are also different for each project.

Inherent in each project are a budget and a schedule. Part of the project process is to perform periodic measurement to track progress.

PARTNERING DEFINED AND WHAT IT MEANS

Partnering is a way of working together and a style of managing that is separate and distinct from industry, organization, or legal entity issues.

> Partnering is defined as a long-term commitment between two or more organizations for the purpose of achieving specific business objectives by maximizing the effectiveness of each participant's resources. This requires changing traditional relationships to a shared culture without regard to organizational boundaries. The relationship is based upon trust, dedication to common goals, and an understanding of each other's individual expectations and values. Expected benefits include improved efficiency and cost effectiveness, increased opportunity for innovation, and the continuous improvement of quality products and services. (Construction Industry Institute, *In Search of Partnering Excellence,* July 1991, page vi.)

SIMILARITIES OF PARTNERING AND ENGINEERING AND CONSTRUCTION PROJECTS

Traditional projects and partnering share the same basic framework elements. The traditional project is a business relationship centered around traditional business practices and approaches and a legal contract. The partnering arrangement, on the other hand, is based on a basic legal contract; however, the partnering concept by its nature goes beyond traditional business practices and is generally not detailed in the legal agreement. The concept and the operating philosophy differences between the traditional project and the partnering alliance are shown in Figure 6.2.

How the traditional cultural practices of management of a project differ from the partnering practices is illustrated in Figure 6.3.

DOES PARTNERING WORK?

The differences between traditional business and partnering concepts and practices are dramatic. In 1991, a Construction Industry Institute study that led to the publication of the report *In Search of Partnering Excellence* summarized the results of a survey of actual partnering alliances from different firms for four key areas. The results indicate that the use of the partnering concept as a management method does offer significant benefits.

Common Elements	Traditional Project	Partnering
A Goal	Complete the Project, within budget and ahead of schedule	A shared goal vision mission statement which includes meeting or exceeding customer expectations, and continuous improvement
Personal Interfacing	Cost Plus = give the client what he wants Lump Sum = Adversarial relationship/ loss of flexibility	Honesty, open communications, focus on problem solution
Duration	The schedule of the project	A long –term commitment
Measurement system	Periodic Budget/Schedule review	Objective / Measurement Matrix with metrics focusing on cost, schedule, safety, and quality (process related variables)
Trust or belief	Belief of competency to completion	Trust of one working for a common goal with no hidden agendas
Reward	Completion of project, sometimes a bonus, a continuation of job	Recognition from openness of sharing of results, self actualization, and sometimes sharing of incentives from performance achievements
Contract	Traditional legal agreement	Traditional legal agreement

Figure 6.2. Concept and the operating philosophy differences between the traditional project and the partnering alliance.

TRADITIONAL VERSUS PARTNERING PRACTICES

Traditional Practices

- Suspicion and distrust; each party wary of the motives of the other
- Each party's goals and objectives, while similar, geared to what is best for them
- Communication structured and guarded
- Single project contracting
- Objectivity limited due to fear of reprisal and lack of continuous improvement opportunity
- Limited access with structured procedures and self-preservation taking priority over total optimization
- Normally limited to project level personnel
- Sharing limited by lack of trust and different objectives
- Routine adversarial relationship for self-protection
- Duplication and/or translation of administrative systems with attendant costs and delays

Partnering Practices

- Mutual trust is the basis for working relationship
- Shared goals and objectives ensure common direction
- Open communication avoids misdirection and bolsters effective working relationships
- Long-term commitment provides opportunity to attain continuous improvement
- Objective critique geared to candid assessment of performance
- Access to each other's organization; sharing of resources
- Total company involvement; commitment from CEO to team members
- Sharing of business plans and strategies
- Absence or minimization of contract terms that create an adversarial environment
- Integration of administrative systems and equipment

Source: Construction Industry Institute.

Figure 6.3

- **Improved ability to respond to changing business conditions**
  Less adversarial 85 percent
  Improved resource planning 85 percent
  Increased openness 82 percent
  Increased trust 78 percent
- **Improved quality and safety**
  Improved safety 90 percent
  Fewer errors 82 percent
  Improved quality 96 percent
- **Reduced cost, schedule, and improved profit (value)**
  Total project cost reduced 8 percent
  Improved contractor profitability 10 percent
  Improvement of schedule 7 percent

- **More effective utilization of resources**

    Engineering cost reduction 10 percent

    Administrative cost reduction 6 percent

    Improved communication and teamwork 82 percent

As the results above indicate, and from our own experience, the partnering concept as a management method does yield a wide range of positive results that have not been demonstrable within the practice of traditional management methods.

Yes, partnering does work and the results are very positive and duplicatable.

Strategic Value of Partnering

The primary purpose of partnering is to improve quality, lower cost, and increase the client's market share and profitability. These are strategic values.

Partnering between an owner and design construct firms is usually structured as a long-term alliance between the two parties. That alliance is built on commitment and trust, and its success is measured in terms of achieving the common goals of both the owner and the design constructor. The design constructor, in a true partnership, will focus on understanding the owner's requirements. The cultures of the design constructor and the owners become one, and the two organizations become indistinguishable as a new single partnering alliance organization with regard to outlook, objectives, initiative, and commitment.

To succeed, the alliance must be based on a mutually beneficial economic arrangement. It also takes adjustment, dedication, mutual support, and belief in the system by the management and employees of the companies. The alliance that is built through strong partnering will improve its organization in every assignment through the attainment of its goals of quality, technical competence, and cost-effectiveness.

The strategic values of a partnering alliance are:

For the Owner

- Improved quality
- Lower total installed cost and investment cost
- Increased market share and profitability
- Improved ability to respond to changing market conditions

For the Design Constructor

- A long-term base of work to stabilize operations
- Strategic relationship building ("I have an investment in you")
- Compensation based on value-added contributions
- Strategic commitment and positioning for future work
- New opportunities and career paths for employees

For Both the Owner and the Design Constructor

- Affiliation with strong and respected national firms known in the business world
- Concentration on the continuous improvement of work systems to provide greater efficiency and enhancement of technical skills
- Better allocation of resources because of the ability to plan on a long-range basis
- A strategic competitive advantage for both owners and design constructors from implementing partnering
- Enhancement of reputations

Partnering Advantages for the Design Constructor

Partnering has an advantage for the design constructor in that it provides a long-term base work load minimizing cyclic variations inherent in the normal bidding process. This arrangement creates a steady and trained workforce and an organization whose efforts are focused on the principles of quality, quality improvement, and productivity management and not on continuous selling and bidding.

Other advantages for the design constructor include

- Increased productivity due to workforce being familiar with the partners' standards, work methodology, procedures, and acceleration of the learning curve;
- Decrease in business development expenses;
- Increased quality of value-added services;
- Less emphasis on multiclient, multiproject selling; and
- Improved morale at the working level and a challenging work environment due to stable base work load.

Partnering Advantages for the Owner

Owners have derived substantial savings and advantages by using partnering. In the past, to handle a varying work load, client organizations were often forced to maintain large engineering and construction management groups at added costs. In many cases, because of the cyclical nature of the work, these groups were not fully utilized. One of the most important advantages to clients has been the elimination or reduction of in-house engineering and construction groups set up to handle plant maintenance, revamp, and capital projects. By having a partnering arrangement, clients have removed themselves from the detailed design and construction business to focus their limited resources on their core business.

Other major advantages for the owner from partnering are:

- Exposure to new methodologies and work processes with resultant savings in costs and higher quality for projects;

- Availability of expertise in a range of specialized technical areas that would be too costly for a client to maintain on permanent staff;
- Elimination of bid preparation and evaluation costs;
- Decrease in cycle time of the project due to elimination of bidding time;
- Involvement of design constructor at an early phase of the project, more over-lap between different phases of the project, and therefore elimination of dupli-cated activities, saving costs, and schedule;
- Exposure to current standards, procedures, and materials;
- Cost savings due to continuous improvement, better quality, less rework, and higher safety standards; and
- Training of owner's personnel in a competitive engineering construction envi-ronment.

Advantages for the Partnering Team Members

The alliance is made up of team members who ultimately make partnering work. Regardless of whether they are from the owner or the design constructor, the partnering environment can be one that offers the dedicated employee a sense of full participation in successful activities providing personal satisfaction and a sense of achievement. It is a continuous process, as shown in Figure 6.4.

In addition, some of the other advantages for employees are:

- Work with some sense of job security because of its duration
- A stated and strong commitment to training and learning, both internally and externally
- The opportunity to make strong personal relationships
- The opportunity to feel a sense of accomplishment and self-actualization by being involved
- An opportunity for immediate recognition and feedback, and in some cases incentive monetary rewards

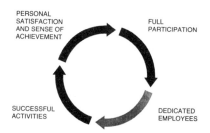

Figure 6.4. The spirit of involvement.

Probably one of the most important advantages is the sense of involvement or self-actualization that the employee feels. This involvement leads individuals to achieve extraordinary levels of performance that have not been seen in an adversarial environment.

Drawbacks of Partnering

Along with all the positives to partnering, there are also drawbacks. From the owner's viewpoint, the drawbacks of partnering can be summarized as follows:

- Overcoming cultural barriers is difficult and stressful.
- The evaluation and assurance of the value received.
- The possible creation of dependencies.
- The possible limiting of competitive strategies.
- The possible development of complacency.
- Internal concerns about job security.

From the design constructor's perspective, the drawbacks can be summarized as follows:

- Overcoming cultural barriers (the old master-slave, owner–design constructor relationship at various organization levels).
- Increased stress resulting from higher client expectations.
- Employee continuity and longevity on the project without affecting individual advancement.
- Owner competitively bidding larger projects even with the partnering relationship.
- Adopting too many of the owner work processes (shortcuts available to a client may put design constructor at risk).
- Learning owner's systems—and not design constructor's—may limit staff development.

Of all these drawbacks, the most important are the cultural barriers. Partnering is culturally opposed to the traditional way of doing business and, unless an ongoing and nurturing culture is present, a partnering alliance will fail. This drawback is overcome by careful selection of prospective alliance members—i.e., those that have the right culture. In addition, the choice of the individual team members who will be on the partnering alliance is also important and has to be done carefully.

Other Important Issues

Two issues that always seem to be in the foreground of partnering alliances today and will probably continue to be issues until the corporate cultures change or become fully receptive and appreciative of the partnering concept are as follows:

- How does the owner determine that it is obtaining the best price from the partnering alliance?
- How does the engineering constructor internally justify the low margin or unprofitable work from the partnering alliance?

The Best Price   Because partnering is a new concept and there are those who are skeptics or who have grown up in the traditional corporate culture, they will always ask, "Are we getting the best price?" This question will continue to arise until the corporate culture moves away from the classic measurement yardstick of price and is realigned to one of overall value. Until then, it is natural to expect this question.

Best price is a classic corporate culture measurement yardstick. It is hard to argue with someone who looks only at numbers and finds the lowest one. However, partnering success is measured differently with a wide set of metrics focused on the five basic areas:

- Overall total installed cost—value,
- Schedule,
- Quality,
- Safety, and
- Owner satisfaction.

There are other intangibles that are harder to measure, such as

- Work process improvements incorporating the best of both organizations and benefiting both;
- The positive openness, problem solving, and trust relationship that develop between the team members;
- Improved profits—bringing plants or services on-line earlier than expected or with better processes or a cheaper capital investment;
- Increased competitive advantage—being able to react to market conditions faster because of trained and knowledgeable staff; and
- Continuous improvement of products and services.

Because, realistically, under the current measurement system there is a need to determine some indicators of best price, the following three areas can be used for benchmarking:

- Comparing the multiplier and fee amounts to the market,
- Comparing the cost and schedule of projects to other internal and similar projects in the industry, and
- Having an outside firm provide measurements for cost and schedule against the industry.

The commercial terms for most partnering agreements have a multiplier, usually a percentage of base wages, to cover engineering and construction overhead, and a fee that is usually thought of as profit. The fee amount is usually so many dollars for each hour worked. Sometimes the fee amount is divided into a base fee and an incentive fee. The multiplier may, in some contracts, be negotiated annually or at the request of either party.

The multiplier in partnering is usually targeted to be set at a level that is not as high as the highest multiplier used in the market nor as low as the lowest multiplier in the market, but as appropriate based on actual data accumulated and presented by both sides in negotiations. The owner uses data from other awards that it has recently made, and the design constructor uses actual cost data. Forward trends are usually considered as part of the analysis.

Because the multiplier is based on market data, it can be considered to be approximately the best price.

The second area where an indicator can be developed is by comparative analysis of the cost and schedules for projects both from the owner internally and externally from the industry. Partnering project costs and schedules can easily be compared in this way to provide further indicators of best price.

The third way we have found to benchmark partnering costs and schedules is to have an outside firm that specializes in quantitative benchmarking make an analysis. This assessment can be used to provide not only a benchmark of best price, but also an assessment of performance of other engineering and construction areas, such as

- Cost deviation of projects,
- Absolute capital costs,
- Engineering construction schedule slip,
- Absolute engineering construction time,
- Absolute start-up time, and
- Absolute early operational performance (second six months of operation).

If these three methods are used, they will provide good indicators that partnering is providing the best price for the owner. As a rule, projects that are run under the partnering concept repeatedly show cost and schedule results better than traditionally run projects.

The Partnering Work May Be Low Margin or Not Profitable By the nature of long-term-alliance partnering work, the multiplier and the fee amounts may be lower than would expected for other competitively bid design construct work. However, each contract is different and client-dependent. The question that frequently arises is why work is being taken that is low margin or for no profit.

The analysis to answer the question can be based on marginal costing. In most alliance partnering arrangements, at the gross margin level, the work is profitable.

It is at the final fully costed operating margin level (the gross margin minus overhead expense and other allocations) where partnering work may appear to be low margin or unprofitable.

The argument internally in many engineering and construction firms is that at a low level of margin or profit, the company should not be doing this kind of work.

At the fully costed operating profit level, there are two possible cases:

- Case one where the operating margin is negative, indicating that the work is not covering the overhead and other allocations, and is unprofitable.
- Case two where the work is break-even or profitable.

Case One   In case one where results at the operating margin level are unprofitable, it is quite possible that other expense elements or allocations are out of line with what should be expected in the industry. For instance, the business development expenses should be lower for partnering work than other work because alliance partnering does not require the usual selling or bidding.

In the area of expense elements, a rule of reasonableness says that if this multiplier is near the going market rate for several other firms, then for them to survive they need to make a profit, so if the partnering work is unprofitable at the operating margin level, then there is a need to review the expense elements to ensure that they are competitive and are not the problem.

Case Two   In case two, where the expense elements and other allocations are reasonable, the partnering work at the operating margin level will be break-even or profitable. In as much as the partnering agreements are long-term and offer steady work, it should be expected that they are by nature lower risk and therefore have lower margins or profits. So it is therefore reasonable to expect break-even or low profit.

In both cases, understanding the impact of marginal costing and absorption of overhead on the profit and loss statement is important. Partnering work pays for or

	Numbers in Thousands		
	COMMERCIAL WORK	**PARTNERING WORK**	**TOTAL**
HOURS	100	100	200
BILLING $/HOUR	$20/HR	$15/HR	
BILLING REVENUE	$2,000	$1,500	$3,500
COST @ $10/HR	1,000	1,000	2,000
GROSS MARGIN	1,000	500	1,500
OVERHEAD	500	500	1,000
PROFIT	$ 500	$ 0	$ 500

Figure 6.5. Comparison of commercial and low-margin work and resulting profitability.

Numbers in Thousands

	COMMERCIAL	COMMERCIAL	COMMERCIAL	COMMERCIAL
	No Partnering	Increase Commercial hours 50%	25% reduction in Overhead	50% reduction in Overhead
REVENUE	$2,000	$3,000	$2,000	$2,000
COST @$10/HR	1,000	1,500	1,000	1,000
GROSS MARGIN	1,000	1,500	1,000	1,000
OVERHEAD	1,000	1,000	750	500
PROFIT	$ 0	$ 500	$ 250	$ 500

Figure 6.6. *Commercial work only, no partnering work.*

absorbs a certain portion of the overhead and that is why, from a financial point of view, it should be done.

Let's take a simple example. Say a firm has a total of 200,000 job hours of work and a million dollars of overhead. Let's next say that one-half the hours are normal commercial contract billings at $20 per hour, and that the other half are billed at the partnership rate of $15 per hour. For simplicity, lets say that the cost per hour is $10 per hour for both commercial and partnering work. (Note, in all cases the rates are fictitious and for illustrative purposes only.) The commercial hours yield $500,000 profit, whereas the partnership hours yield very little or no profit (see Figure 6.5).

Someone could say let's just stop this low margin or unprofitable work. As a result, the firm abandons its partnering work and ends up with 100,000 hours of commercial work billed at $20 per hour and overhead of at least a million dollars. The firm is now at best break-even, and the $500,000 of profit has evaporated.

The point of low margin or baseload partnering work is that it provides gross margin that can absorb part of the overhead. To return to its former level of profitability, the firm has to either increase the number of hours billed by half again at the same margin, or it must cut overhead by half (see Figure 6.6).

One might argue that by keeping the base hours the firm is neglecting markets that it should be entering. There is always the big "if": "If" you get the job. However, and with strong management, they should be pursuing new markets in addition to partnering base work load.

The partnering work may be low margin or unprofitable, but it does absorb overhead, and it is a bird in the hand.

7

Benefits for the Project Commercial Professionals

In the engineering and construction industry today, the project commercial professional has the responsibility for the key areas of administration, contract compliance, transaction accounting, and management reporting. These areas are traditionally found in the engineering and construction environment. Under the partnering concept with the inherent focus on client needs and continuous improvement, some evolution is taking place.

As a member of the partnering team, the commercial professional participates with full empowerment. The accounting function may or may not be resident on the project. The positioning of the group is usually a function of the group's reimbursability in the contract. When the group is reimbursable by the contract, it is usually located on the project; when it is not, it is usually separate. In some cases, position may be dependent on the size of the partnering alliance.

The two areas where we have seen an evolution of the commercial relationship have been the reduced invoice inspection and verification processes. By modifying our accounting work processes to meet client needs, the clients' accounting and internal audit functions have been reduced, which resulted in savings. We have found that in the financial area, the partnering alliance benefits most from the work process improvement that takes place.

By modifying our computer systems to meet client needs or by actually using client systems, we have facilitated the accounting and reporting processes, thus saving money. This has led to a more focused effort from our accounting staff as well as the owner's. On one project where our accounting group was using part of the owner's systems and had gained a good knowledge of the rest of the system because of the required interfaces, the financial person was called on to help the owner because the owner's project accounting group had been downsized and our group was the most knowledgeable about the owner's systems.

Another area where we have seen benefits is in the area of imprest accounts. Many of our clients provide cash in imprest accounts, resulting in no accounts receivable between the partnering members. Cash requirements are determined and requested in advance from the owner. The owner deposits the money and project management uses it on a current basis.

8

Don't We Partner Already?

If you are in a work situation where all the key players have a shared vision of what extraordinary accomplishment would be; and if your group gauges what to do, when to do it, and how to do it first and foremost by how it will affect the final result; and if your group puts the pursuit of extraordinary accomplishment ahead of unreasonable policies, procedures, traditions, and protocol, then you are partnering (see Figure 8.1). With such high expectations set out for partnering, few work groups are partnering, and a fair number of so-called partnering groups may not really be partnering either.

In another sense we could view partnering as being at the high end of a continuum, and we could view partnering as a matter of degree. At the bottom of this scale we would put dysfunctional work situations. Moving up the scale we would put marginal performance groups with obvious problems. Further up the scale we would put groups that are working well and doing a good job. Quite high on the scale we would put work situations that have mastered good teaming skills and are serious about high performance; at this level we find esteem and pride playing key roles in motivating behaviors and performance above and beyond the minimum required. The esteem and pride measures of the group generally require the admiration of others in order to work well, even if it is just mutual admiration among the members. At the very top of the scale we would put work situations where both individuals and their groups have transcended to a level of self-actualization toward extraordinary accomplishment. At this level the motivation is inner-based and self-sustaining. At the partnering level we find that the challenges and rewards are almost completely internalized. Using this continuum we can place all work groups somewhere on the scale and say that they are all partnering to some extent.

Agreement			Partnering exists in a work situation if...
+2	+1	-1	
H	M	L	All the key players have a shared vision of what extraordinary accomplishment would be,
H	M	L	The group gauges what to do, when to do it, and how to do it by first and foremost on how it will affect the final result, and...
H	M	L	The group puts the pursuit of extraordinary accomplishment ahead of unreasonable policies, procedures, traditions and protocol.
Total Score ✍			GOOD PARTNERING WORK ≥ 3

Figure 8.1. *Testing for partnering at work.*

However, in all fairness to partnering, we believe it deserves its own scale. A single-minded focus on extraordinary accomplishment sets partnering distinctly apart from other working relationships and styles. If this single-minded focus on extraordinary accomplishment transcends interdepartment and intercompany lines, then we have full, classic partnering.

The drive for extraordinary accomplishment necessitates extraordinary behaviors, and we can look for these behaviors as indicators of partnering. Especially essential is open and straightforward communication. Following common focus and open communication will come knowing where you stand with each other, and from that comes camaraderie and trust.

Is a group partnering already? To the extent that a group scores true on the following questions, they are partnering:

1. The group is clearly engaged in a common pursuit of extraordinary accomplishment.
2. The group is unified in challenging what it finds to be unreasonable policies, procedures, traditions, or protocols that significantly restrict their accomplishments.
3. Group participants enthusiastically practice open and straightforward communication; there are no signs of guarded participation or hidden agendas.
4. Individuals within the group freely exchange concerns, ideas, and criticisms, and they appear to fully trust each other (see Figure 8.2).

With only a limited amount of observation of a work group, an observer should be able to answer the above questions, even if they have to use shades of true and false (absolutely true, true, probably true, probably not true). If the above conditions hold essentially true in a work setting, and especially if they hold true while crossing departmental and organizational boundaries, then a form of partnering is surely at work.

Agreement			A group is partnering to the extent that it demonstrates the following behaviors...
+2	+1	-1	
H	M	L	Clearly engaged in a common pursuit of extraordinary accomplishment,
H	M	L	Unified in challenging what it finds to be unreasonable policies, procedures, traditions or protocol that significantly restricts their accomplishments,
H	M	L	Participants enthusiastically practice open and straightforward communication, there are no signs of guarded participation or hidden agendas,
H	M	L	Individuals within the group freely exchange concerns, ideas and criticisms; and...
H	M	L	They appear to fully trust each other.
Total Score ☜			**GROUP PARTNERING ≥ 6**

Figure 8.2. *Testing for group partnering.*

If we redefine partnering and include groups that are working well together for mutual benefit, then we can say that many groups are already partnering. To some extent this would seem to dilute the aura of partnering. However, an important advantage to putting more work groups on a partnering scale is that they can appreciate that much of what they are doing is in line with partnering, and the pursuit of partnering becomes a growth and improvement issue that builds on what exists. Looked at this way, partnering is not a major change, it is a growth opportunity. If a work group is already engaged in a substantial amount of the work they will be partnering for, then introducing partnering as an improvement to the next higher level of work style is appropriate. If the work group is in the early formation stages, then pulling the group together and launching the partnering effort from ground zero is likely to be more appropriate. As with partnering itself, flexibility in best serving extraordinary accomplishment is the key in launching partnering, not setting and following a structured program.

9

When Does Partnering Start?

For those who will not be directly involved in the partnering group, the question of when partnering starts has more to do with the timing relative to the project. Briefly, partnering works best when it starts as soon as the concept for a project is introduced. Instituting partnering with representatives of all major parties at the very beginning allows for the best options to emerge early enough to be incorporated. As the project moves forward from its inception, major options and choices are gradually reduced. Significant reorientations of the project that might improve life-cycle economics simply cannot be entertained after a certain point. Reengineering material flow is impractical later in the project. Major breakthrough thinking and cooperation must come early enough in the project to be incorporated, so it is in the owner's best interests to start partnering as soon as possible. However, introducing partnering late is generally better than not introducing it at all. Even partnering introduced at the very end of a project can significantly improve the handling of disputes and punch lists, so better late than never.

For those who will be direct participants in a partnering group, when partnering starts involves a different set of issues. For example, if we say that partnering starts with the first meeting of the group, we would be attributing partnering to accomplishing the process. Partnering is not about doing a process or even doing it well. Partnering is about accomplishing the extraordinary.

Having the first meeting and pumping out a mission statement, group principles, and discussing the essentials of partnering are important to partnering, but they are only "get ready" steps. While we do not want to credit a group with starting partnering just because they completed an important meeting, we also do not want to have to wait until the project is practically over before we gauge whether or not partnering is alive and well. Beliefs and behaviors are the foundation of partnering. Central to accomplishing the extraordinary is open and straight-forward communication, freely exchanging ideas, concerns, and constructive criticisms, and working through constructive confrontations. When we see these behaviors at play, especially in interdepartmental and interorganizational settings, we know that partnering has started. When the discourse in meetings moves beyond guarded positioning and following protocol, partnering has started. When individual differences of opinion are not emotional, defensive, or win-lose exercises, but are logically evaluated for what will best serve extraordinary accomplishment, then partnering has started.

Partnering does not commonly take off with a bang. When building good team skills and camaraderie, many of the new participants think partnering starts with the first meeting. But seasoned partnering individuals and facilitators usually see it differently. During the first workshop, a few in the group start climbing on board rather quickly, but the others are usually sitting back and waiting to see what happens. The group does go through the motions, but only a minority really step from the old path and walk on the new side. The others pretend but their discomfort and restraint are usually fairly obvious.

When a few participants use open, straightforward communication to work through some tough issues, and once they work their way through one or two constructive confrontations, others begin to understand what partnering is about and believe that the new behaviors might work after all. But the majority still tend to hold back just in case all this openness backfires and those who do open up end up regretting it. Waiting to see how it works and whether it is a smart thing to do is the norm. Then, at some point, the flood gates open and almost everyone jumps in. With some groups it only takes perhaps a fourth of the participants to join in and open the gates. In other groups it takes a third or a half.

But after several hours together, it generally happens rather quickly. Typically the group is getting more relaxed, more friendly, more comfortable, and somewhat convinced, and then a sensitive confrontation or issue pops up. Most draw back and wait for the new partnering system to break down or blow up. But it does not fail. It works. With a success on top of a half faith, there is almost an audible sigh of relief from the group and the guardedness and old ways begin to fall by the wayside. The threshold of partnering was reached, a tough situation was successfully worked out, faith in partnering increased, and the group crossed the line into real partnering. Certainly there are always a few stragglers, and some never make it. But when the work culture is comfortable enough that those who buy into partnering can get on with it in spite of the stragglers, then partnering has started and is on its way.

To be able to determine exactly when partnering has started would require mind-melding with the participants. When the majority of participants is thinking first of extraordinary accomplishment before worrying about job security, job descriptions, egos, interpersonal relationships, policies, protocol, and the like, then they are thinking as they must for partnering. When the little voices in their heads start frantically bringing up new ideas and approaches and asking "Why not?," then they are partnering. When they have a propensity to not avoid difficult issues but to act on such things that could reduce constraints or improve results, then they are partnering.

Partnering shows up in the work and in the final results. Partnering exists in the minds of the participants. We see partnering behaviors, interpersonal dynamics, group dynamics, and they are all indicators of partnering. Partnering is in the thoughts and beliefs of the participants. When the inner thoughts manifest themselves in outward behaviors, we can tell that partnering is taking hold.

10

Benefits for the Owner

Because EPC partnering's objective is extraordinary accomplishment and because extraordinary accomplishment is defined as the best life-cycle outcome for the owner, the owner is the major benefactor of partnering. Participants other than the owner tend to benefit to some extent, but not to the degree that the owners benefit.

FROM COMPLIANCE TO OPTIMIZATION

Partnering changes the work process from one of compliance to one of optimization. Whereas non-partnering groups are driven to meet the owner's needs by the best performance against the contract, partnering groups are openly driven to meet the owner's needs by the best final result possible under the circumstances, regardless of the original plans. In non-partnering situations a considerable amount of time is spent figuring out how best to meet the requirements of the contract. In partnering groups some of that time is redirected to making the outcome better for the owner, sometimes even in spite of the contract. In non-partnering situations each person strives to do their job and do it well. In partnering groups everyone's job is to work together in the best way possible to accomplish the extraordinary. Time formerly used for policies, position descriptions, contracts, requirements, and watching to see if the other players are doing their work is now used to pursue the group's ideas for best results. To achieve extraordinary accomplishment partnering will certainly meet its legal, ethical, and contractual requirements, but partnering generates an aura of cooperation and enthusiasm to do much more than just what is required.

Partnering's open cooperation among all major past, present, and future players contributes much to eventual best value for the owner. Because of partnering's broad cooperative effort and high-level view of the work, the owner's schedule and cost are not the only areas that receive attention. On the owner's behalf the group also seeks to improve life-cycle economics, public relations, and long-term corporate improvements that start with the project at hand but favorably affect owner operations as well. Certainly, partnering isn't the only approach to gaining benefits from open cooperation. Owners do an excellent job of developing their projects by involving all the major parties. However, partnering takes that practice a step further. Once the owner turns the project over to the EPC contractor, the contractor continues that wide involvement to spot problems and opportunities all the way through the project and right up until the project is released back to the owner and its operations and maintenance crews. Partnering ensures that everyone does their job, but it also ensures that participants deliberately pause from time to time to exchange ideas and concerns and work out better alternatives and plans.

THE LONG-TERM VIEW AND CONSTRUCTABILITY

For most projects partnering is able to increase the long-term value to the owner. However, for some projects, even with partnering, there is little difference between original plan and final result. There are those cases where the original concept and plan were seriously flawed and partnering's open approach was able to identify and resolve the shortcomings and deliver the final result that was intended. Converting a potential disaster into a successful project can certainly be considered an extraordinary accomplishment. In all these cases the major benefactor is the owner.

Under standard EPC business practices, building constructability into the front end of a project and keeping a sharp focus on it throughout the project is a special project or activity. The constructability effort needs to be defined, managed, and inspected. Testing for constructability quality and effectiveness as a separate exercise can get to be a considerable effort. However, in partnering the focus is on extraordinary accomplishment, and constructability is one of the built-in issues that is important to the final result. Partnering's own management cycles (planning, doing, checking, and adapting) ensure that the constructability effort is neither too little nor too much, but that it is integrated with other activities essential to producing the best result. Traditional work groups can plan and carry out constructability reviews, and their focus is often more on the review than the final result. Partnering's focus is on how constructability best plays into extraordinary accomplishment. The difference in perspective is important to the final outcome.

RESOLVING REDESIGN AND REWORK PROBLEMS

Owners sometimes have to accept schedule and cost adjustments for critical redesign and rework problems with their projects. Standard EPC contracting's view is that the owner is responsible for the design base, and EPC picks up from there. If the EPC contractor in executing the contract gets far enough downstream before a design-based problem crops up, and they were doing their job, the owner is in a position of having to accept whatever contract and budget changes are necessary to get the project back on track. The cost will be absorbed by the owner unless the EPC contractor technically failed in some way. The EPC contractor then reacts to the situation and the project moves forward.

Partnering helps to protect the owner from such unfortunate changes. Upon formation the partnering group immediately takes on board the right mix of people and accepts the challenge of thinking and rethinking the project in the owner's best interests. Problems obviously still arise, but because of partnering the problems are usually recognized sooner and resolved faster. Partnering goes for the best solution as quickly as possible with minimal regard for organizational or contractual barriers. Resolution in the best interest of the final result is the primary focus, not contractual compliance or dispute development and resolution. Time, creativity, and effort are largely redirected to better uses—i.e., better using time that might otherwise be spent defining the problem in great detail, studying who is responsible for it, studying how the problem is affecting other aspects of the project, and sorting out who should do what and who should pay.

USING TRUST AND HONESTY TO CONTROL CONTRACTORS

Along with the effort by the EPC contractor to improve the owner's lot comes a concern. Might the EPC contractor's efforts to improve the project be a front for improving the contractor's position? Through partnering, isn't the contractor in a good position to use smoke and mirrors to dangle pipe dreams in front of the owner and end up with an easier or larger project that benefits the contractor's coffers more than the owner's? To the extent that an owner might abdicate a project to an EPC contractor and have insufficient involvement, then such an exposure is real. But the owner should be a key player in the partnering group even if the EPC contractor shoulders the majority of the burden. As long as the owner participants are genuinely engaged in open, straightforward communication, they will have a great opportunity to know what is evolving and will be able to contribute to the process. They should raise possible conflict-of-interest issues with the EPC contractor when they arise, and the EPC contractor will be obliged to openly work through the issues with the owner participants.

The work style and fundamental nature of partnering are built on trust and honesty. The likelihood of an EPC contractor taking advantage of an owner is lower in an open partnering style than in a traditional contracting situation. When executed even minimally, partnering transcends what is good for each party and focuses on what is good for the owner's long-term interests. We know that in the real world there will be those who will try to take advantage of the situation. But in the partnering world, they are less likely to be successful. Through regular self-assessment workshops, the partnering group will usually sort out those participants with personal agendas and subversive behaviors and have them replaced. Partnering regularly tests itself for the right membership and deals directly with problems it finds. Such direct assessment and action are part of the partnering management cycle, and, once more, the owner is the primary beneficiary.

LUMP-SUM FIXED-PRICE WORK

Lump-sum fixed-price projects bring a special set of interesting issues to the partnering table. If through partnering the EPC contractor significantly trims costs and schedule, who benefits then? The EPC contractor usually wins an early completion bonus, loses work for their resources, and earns additional profits. If the contractor has new work to move its resources to, it gains handsomely. If not, the benefits of trimming costs and completing earlier may not be very attractive in the larger view that includes idle resources. In lump-sum work the owner may not be the primary beneficiary of partnering. To the extent that the team heads off problems and comes up with new and better solutions, the owner may end up with a better project than it contracted for, and the alternative dispute resolution gains may be another big plus for the owner. Additionally, the early completion bonuses are usually much less than the operating benefits from starting early, so the owner can gain some operational profitability too. The cost and effort incurred in participating in partnering on lump-sum projects by owners still represents a good investment. If the biggest gains on fixed-price work are cost reduction and early completion, the EPC contractor may gain the most from a partnering effort.

Partnering can profit owners in a multitude of small ways too. The owner's staff who participate in partnering can grow dramatically in their teaming and interpersonal skills. Being a part of such an open and assertive group helps all the participants, and getting comfortable with a no-holds-barred pursuit of extraordinary accomplishment is a work style that the participants can use for the rest of their careers, not to mention their personal relationships. Owner participation on the partnering team also exposes participants to the kinds of thinking and changes that the EPC contractor develops to reduce costs and accelerate completion. These insights can be useful to the owner in planning future projects.

There are more reasons why owners should participate in partnering on lump-sum projects. By participating in the partnering effort, the owner may become sufficiently confident in the EPC contractor to consider sending additional work their way without the usual cost of developing RFPs (requests for proposal) and negotiating contracts. Being a good participant on the EPC's partnering effort can score public relations points, and if regulators are a part of the partnering effort, the owner can strengthen those relationships that can help in other areas of the owner's operations. Owners with a reputation for partnering are in a position to leverage that reputation in acquiring talent that might otherwise go to the competition. Issues will often come up during the partnering effort that may not be significant enough for the EPC contractor to act on, but the owner may be in a position to use the information to its own advantage. Such ideas might include environmental issues, intervener positions, supply or demand market changes, supplier issues, and ideas for long-term operations and maintenance. The EPC contractor might also momentarily consider a quality improvement above and beyond the contract but choose to move on with the existing specifications. If the owner is involved in the partnering, it might give the idea further analysis. If warranted, it can then come back to the contractor with a change order to capitalize on the idea.

SHOULDERING POSSIBLE CRITICISMS

What is in it for the owner is not all good news. Along with the good, there will be some bad. Especially in the beginning of partnering, the owner can come under criticism for stepping away from standard EPC practices. Stockholders, investors, customers, insurers, bankers, joint venture partners, unions, stock analysts, regulators, and just about anyone else can express concern that partnering might not be right for the work at hand, or that partnering may not be in their or the owner's best interests.

As with nearly all new approaches or technologies, dealing with these criticisms requires careful listening (especially since good points may emerge from time to time), clarifying the issues, responding with appropriate information and insights, and assuring those concerned that the owner will keep them apprised or involved as needed. Some will not be satisfied with any response they get, so the owner may from time to time simply take the position that partnering is seen as the best option and until proven otherwise it will stay. Fortunately with partnering, most of the interest and press is favorable, so owners should face minimal opposition. Still, to head off concerns about partnering, owners would do well to provide regular updates to all interested parties and to especially get the good news out when partnering seems to be doing better than might have been the case under traditional contracting arrangements.

Because partnering's goal is extraordinary accomplishment and the best result under given circumstances, the group at times will bypass normal protocol to speed things along for a faster, better outcome. Owners can be criticized for having a partnering group that seems to be out of control if it appears that the owner doesn't know what is going on. Partnering's blunt pursuit of results can make many people feel uncomfortable and threatened. Owners would do well to monitor partnering groups to ensure that they do not get too rambunctious and rock the boat more than is needed. Partnering groups are responsible for being tactful in their dealings outside the group, but the chase for extraordinary accomplishment can get them overenthused at times, and their tactfulness may be less than appropriate. In most cases the offense is taken not so much by the people they deal with, but by supervisors who feel left out or jilted by the partnering group's direct contacts and unreviewed actions.

TRANSFERANCE TO ALLIANCES

Owners can gain valuable experience through partnering on EPC projects that can be transferred and used in intercompany alliances. Partnering at the individual level and in EPC work provides an excellent training and development experience for those who might be sent on to work in larger scale alliances. The partnering background can be especially valuable in international alliances where major cultural differences and lack of trust can derail standard business practices. Partnering's interpersonal training and practices provide an excellent platform for moving past cultural and diversity issues. Within a partnering work setting, diversity of any kind is simply a factor to be addressed on the way to extraordinary accomplishment.

If gender, race, belief, disabilities, background, physical traits, or any other individual issues interfere with accomplishment, they are raised and resolved along legal and ethical guidelines according to what is best for the final result. Personal biases and preferences are not allowed to play into the resolution except to the extent that they have a material impact on the business at hand. Through partnering training, diversity is just business is business. No hidden agendas, no personal agendas, nothing is taboo, and good people are expected to work through their issues and get back on track.

ORGANIZATIONAL IMPLICATIONS

As owners move away from standard EPC arrangements and more into partnering arrangements, they will find that less time is needed on rework, redesign, auditing, inspecting, testing, checking, responding to inspection responses, following up on inspections and reports, meetings to discuss inspections and responses, and so on. Most of the talented people who used to do those things can be reassigned to identify and resolve problems and research opportunities. Instead of finding and reporting problems, they are responsible for finding and solving problems, and chipping in on anything that will help the project. Some owners can take advantage of the reduced workload and reduce staff and contractors accordingly.

Partnering can bring a whole new world to owner engineering departments. Under the new working relations the owner engineers sometimes find that they are a part of the EPC team assigned to review and improve anything and everything. Partnering groups do not tiptoe around questions and issues, especially when it comes to owner specifications, drawings, and plans. Without any particular ceremony they usually just lay issues and questions out on the table for discussion and resolution. Some owner engineers are threatened, intimidated, or angered by such an affront to their work. When partnering groups have extensive meetings with those who are not directly in the partnering effort, such as the owner's engineers, special workshops for the non-partnering people should be held to help them understand partnering's style and how to work with it.

After working hard to develop final materials for the EPC contractor, it can also be unsettling to the owner's engineers to find that their work is not cast in stone. Their work is continually challenged to improve it right up until the final stages. Fast-track engineering and construction has been around for years, but partnering pushes fast-tracking right back to original specifications and drawings. In place of formal information requests and exchanges with the owner's engineers, the partnering group usually just walks through the door and starts asking "Why?" and "Why not?" If that causes the owner engineering staff a problem, it usually does not last long. Those who can't adjust are simply taken out of the picture one way or another. However, just about everyone gets a kick out of being a part of breakthroughs and extraordinary accomplishment, even if it means having their work redone.

In the past owners have trusted in their inspections and controls to ensure a satisfactory project. Shifting to trusting people, especially outside contractors, is a major challenge. Being a part of the training and development is important, being a part of the process is important, and being a part of the partnering management cycle is important. Just as partnering participants have to build trust in order to function properly, owners need to build trust in partnering in order to fulfill their role.

Owners who take partnering on board generally realize many improvements, some on the projects and some that influence other areas of their operations. To the extent that the owners begin finding improved ways to do their business, they may face an uncomfortable reality of eliminating positions in order to capture the benefits. Rolling in partnering and then rolling out no-longer-needed positions is an important way of realizing the benefits of partnering, but not necessarily an easy sell to those affected. The partnering group will be sensitive to the implications of its improvements, but at the same time it must have a very business-is-business attitude. If improvements are evident, they should be recognized. If management chooses not to take advantage of the improvements, that is reality. Owners may find such an open pursuit of improvements unnerving. Wanting improvements is one thing; terminating positions or making the changes is another.

In some cases transfers and attrition can be used to capture the reduced position benefits, but sometimes terminations are the only way to capture the benefits. Keeping unnecessary positions is a detriment to extraordinary accomplishment and an affront to the partnering effort. Unfortunately, not eliminating unnecessary positions is quite common in many organizations. Fortunately, partnering accepts reality. If they have made their case as well as they can, and they have not been successful in persuading management to act, they should back off and get back on track for extraordinary performance. Wasting time on benefits that are being turned down is counterproductive to the larger calling. For partnering it is all in a day's work. For management that won't implement benefits, it may be a wake-up call. Maybe the manager who won't act on benefits today will at some point in the future be seen as something to be resolved on the way to extraordinary accomplishment.

And now an important qualifier. Just because the partnering group thinks it is on to a great benefit doesn't make it so. Remember, from time to time partnering will be in error. If management doesn't act on the partnering group's recommendation because they believe the group is in error, then management is doing its job. If management does not act on the partnering group's recommendations because they are "chicken," then that is a bird of a different feather.

Many of partnering's breakthroughs do not involve eliminating positions, but some do. When partnering takes a fresh look at how business is done and how companies work together, it happens that some positions, and even whole departments, become redundant. Heretofore important and busy people are quite suddenly not needed. As the lights start going out, these displaced people may at first deny what is happening. They may be dismayed and be waiting for someone to tell them that their careers are really okay. But they often get angry, and then they usually start fighting back. And not only do they start fighting the elimination of their positions, others around them join in the struggle to save the positions, even though their positions are not threatened. Some join in because they do not think it is fair; others join in just because they think it is the right political thing to do.

Partnering is charged with doing what it can to pull all these people into the process to be a part of the solution and to push forward in whatever direction is to their best advantage. However, partnering has that single-minded focus on extraordinary accomplishment toward their mission. The partnering group will be able to work toward implementation to the extent that (1) it is charged to do so or (2) it is an effort that will promote accomplishing the extraordinary.

Where the positions or departments do not report directly to a member of the partnering group, dealing with displaced positions is not partnering's responsibility. Identifying business improvements is the group's responsibility, not forcing themselves in where they do not belong. To take full advantage of partnering group ideas and capture the secondary benefits from the group's efforts, executives associated or approached by the group should stay tuned in to what recommendations the group is offering and cooperate with the teams whenever it seems appropriate. Partnering will work hard at implementing recommendations within its realm. Associated executives will need to take responsibility for recommendations made outside the group's realm. Not pursuing secondary benefits from partnering can be a significant loss.

SHATTERING TRADITIONS

Partnering attempts to shatter some of our long-standing rituals. For example, owners tend to lean hard on EPC contractors to squeeze as much out of them as possible. When the contractor cuts too many corners and stumbles, the owner inspection forces come down hard on them and force the contractors to make things right or pay damages. The owner usually wins and the contractor usually loses. But at the same time, some contractors cut corners, stumble, and do not get caught. The contractors also from time to time realize that hours and days could be cut off of the work, but that would mean fewer hours for their people and fewer dollars in the coffers. It is no wonder that an adversarial relationship evolved.

By getting everyone on the same side of the table and aiming at the same goal, partnering essentially eliminates these outdated games. But wait, you say, if the owners usually won under the old rules, why would they stop the game? (Actually a fair number of owners have not stopped.)

OWNERS GAIN THE MOST

The problem with these old-style games is that the owners only won a few points on a couple of scoreboards. They saved or recovered costs against a set of specifications and contract clauses, which could actually be several million dollars, but they lost more. From a life-cycle economic standpoint, total installed costs may have been lower but downstream costs were often slightly higher than what they could have been. Slightly higher costs over the life of a project usually total more than near-term cost savings. Projects that could have been completed earlier, even at the same cost, were just completed on time. The lost days of operation usually amount to more than the costs that were saved under the old style of pressure. And projects that could have been improved for better throughput or efficiency were not, and over the life of a project the marginal return can be a rather large sum. Partnering dismantles the management and control systems that hindered a better EPC project for the owner. It is a new game and, played right, the owners still win more often, but they win bigger. And as long as there is some sharing of the marginal improvement in life-cycle benefits to the owner, the contractor does not lose and may even win a little too.

Are we saying here that the owners receive the most benefit from EPC partnering? Experience to date seems to suggest that is the case. Certainly EPC contractors who are good at partnering are also fairly good at negotiating contracts that will reward, at least to some extent, their contributions to the owner's best interests. But the return to EPC contractors does appear to be considerably less than the relative return to owners.

Some Gains for the Contractor What may improve the balance somewhat is that EPC contractors who are perfecting partnering and are adding greater value to the owner's bottom line are more and more being given opportunities for noncompetitive follow-on work if they can qualify for it and want it. When the initial partnering work and the noncompetitive follow-on work are reasonably profitable, the EPC contractor at least keeps its forces billable and can continue to build expertise and credentials. Even when the profits from partnering projects are disappointing to EPC contractors and there is no follow-on work, their reputation for exceeding owner expectations can help them land other work. One additional benefit that also seems to be emerging for EPC contractors who are good at partnering is that they are continuing to hone their performance and may eventually outlive EPC contractors who are not partnering and are not honing their performance as well as the partnering EPC firms.

Over the next decade we may see the best firms and those who do a fair job of partnering survive, and we may see a great number of non-partnering contractors disappear. When we look at the industry in another 10 years, we may find that the majority of EPC contractors will be proficient in partnering, and essentially none of those eliminated in the next 10 years will have been proficient in partnering. Separate from the obvious benefits for the owners, the internal and external benefits of partnering excellence may simply be a good survival strategy for EPC contractors.

WHY SHOULD OWNERS SHARE THE BENEFITS?

When EPC projects start, the playing field and scoreboards are usually pretty clear. Here is the project, here is the contract, do a good job, and here is your fee (or margin on fixed price). The contractor gets what it deserves, the owner gets what it deserves, the parties settle any outstanding claims one way or another, and the game is over. What happens when we bring partnering on board? If the owner doesn't participate, the contractor uses partnering to maximize its own satisfaction, which includes to at least some extent a satisfied owner. If the owner joins in and the partnering group comes up with some impressive cost and schedule gains, the owner wins and the contractor hopefully isn't any worse off. But clearly the contractor isn't overly excited with the outcome when the owner walks away with all the loot.

If the owner joins in the partnering and offers to share the booty of major breakthroughs, then we have a full-blown game. Everyone wants to find the absolute best extraordinary accomplishment. Everyone has the potential to be a winner. However, a fair number of owners are questioning why they should give away anything beyond a few token marbles. Their argument goes like this: If the owner's sharing of extra marbles is minimal, the EPC partnering group is still going to do a great job. Let us say they shave off two weeks and save $10 million. That was their job in the first place and, after all, business is business. Their argument continues that if the owner agrees to share the marbles from the breakthroughs, there won't be all that many more breakthroughs anyway, and the owners could have avoided coming up with a sharing program, much less trying to apply it. For example, the added motivation might result in shaving off three weeks and $15 million. By the time you dole out the shares, there is little difference to the owner. Believing this, the owner opts to keep the marbles, say thank you, and offer the good contractor some additional noncompetitive work as long as its performance stays high. In several segments of the industry, access to additional noncompetitive work is a valuable outcome of partnering, so we can't really knock it. But the point we are going to make later on is that the owners would likely be even better off if they entertained a more significant sharing of partnering's breakthroughs.

In the end the owner will take whatever path looks to be in its best interest. If a partnering group can make a case to an owner that knowing that the extra bonus is available will allow them to put more into the project, then maybe the breakthroughs might add up to four weeks and $20 million, and the owner would probably get to keep $15 million. If true, the owners would be better off for sharing more of the marbles. Experienced people can rough out what they think will happen with a specific project under a number of scenarios, but it is anyone's guess what will really happen.

As we discuss elsewhere, we happen to believe that if the marbles aren't shared, partnering will produce notably less benefit to the owner than if arrangements are made to share. Even to the inexperienced, it makes good intuitive sense that if the owner and EPC contractor are working together to achieve the extraordinary, and if both parties stand to be handsomely rewarded for their input and cooperation, then what chance is there that the result would be disappointing? If it works and partnering blows the socks off the plan, everyone wins. If extraordinary results do not materialize, the players are no worse off than before. Low down side, high up side; that's a good investment in anyone's book.

SHARING BENEFITS AND RISKS

How to share the marbles can be a really troublesome issue. In some cases the industry has well-established programs. Dollars-per-day bonuses and penalties can be applied to the agreed completion date. Dollars-per-efficiency bonuses and penalties can be applied to design specifications. Cost plus a variable fee based on percent of budget can be used. The EPC contractor's fee can also be based on a long-term profit-sharing plan to encourage life-cycle economic considerations in the EPC work. So, proven and common tools are available, and we have enough experience with them to plan ahead for the usual kinks and quirks.

Can these established bonus and penalty plans work with partnering? Sure they can. Do we need anything else? Maybe. Some breakthroughs require extensive reengineering and result in a significantly different total installed cost and future operating margin. Some breakthroughs have more to do with long-term liability or environmental concerns. What then? Such breakthroughs either tend to blow the standard bonus plan out of the water or don't fit in, as in how to value a long-term environmental improvement that has essentially no short-term impact.

As a matter of course partnering is geared to finding solutions that lead to extraordinary accomplishment. Finding blame and filing claims just do not happen unless everything else has failed. When standard partnering can't seem to handle a claim or damage situation, they typically turn to a special SWAT team of one form or another. The special team seeks a resolution between friendly forces before any big dollar wars get started. Neither the owner nor the EPC contractor ever seems to really win anything through litigation. The final outcome is usually just a matter of who is less worse off. Already in partnering we are seeing impressive claim avoidance and resolution programs that head off problems with solutions and quickly get the parties back on track. Perhaps a similar process can be used from time to time to iron out equitable sharing of unusual risks and rewards.

The risk-and-reward SWAT team would take all things into consideration and work out an equitable balancing between the parties, much like they do with claims and damages issues. We haven't seen such a formalized program yet, but we think it would be a good idea and we are looking forward to setting up our first risk-and-reward SWAT team.

If we can get a risk-and-reward SWAT team launched, we have another job for it, partnering public relations. Partnering needs to show a profit or other demonstrable value just as any other business endeavor must. Most of us in the industry are still running on hyped-up stories and faith. The Construction Industry Institute and the U.S. Army Corps of Engineers have collected and demonstrated the value of partnering, but we need to do more. Perhaps it is the risk-and-reward SWAT team, perhaps it is someone else, but each partnering group should get an auditor. The auditor can track marginal costs of partnering versus marginal gains from partnering and keep the group and key executives well informed as to how things are going. If they are going well the results can be publicized and help the group along. If things aren't going well they need to be resolved one way or another. If partnering is in trouble and can be turned around, good. If changing some of the partnering participants is the key, do so. If partnering is digging a deeper and deeper hole and there is no realistic hope, shut it down. Partnering's own single-minded focus on extraordinary accomplishment should drive it to fix itself or get out of the way. If it doesn't, the partnering auditor has a responsibility to step in and draw attention to the situation. With good people in a good setting with good facilitation, partnering can hardly miss. But if it does miss, it needs to be shut down with a minimum of damage. Bad partnering is not good for the industry, and it is not good for partnering.

THORNY LIABILITY ISSUES

While we are beating up partnering, we might just as well throw in some more bad news. It is true that effective partnering is making its mark on many fronts in the industry. But it is also making a few waves and raising some thorny issues. When the EPC contractor starts acting in the best interests of the owner in the pursuit of extraordinary accomplishment, sometimes they influence the project in ways that used to belong to the owner. At what point does the EPC contractor start picking up a share of the owner's liability? If the EPC contractor just does what it is told, the liability picture is fairly clear. But as the EPC contractor spends more time on the owner's side of the table, what then? The reverse is also true. In partnering, owners don't just tell EPC contractors what to do and hold them responsible for meeting specifications and schedule. In partnering, the owner's engineers spend far less time inspecting and much more time figuring out better ways to get the job done. At what point does the owner begin to pick up some of the EPC contractor's responsibility for the final product and the EPC liability that goes with it? And for a final twist in partnering, regulatory agencies are being asked to put their time and expertise into helping the project comply rather than inspecting it and processing complaints against it. The regulators are actually help-ing compliance more by lending a hand than inspecting and reprimanding, and it costs the taxpayers less, but they lose their arm's-length enforcer role. Will the public let them join in the execution and then trust that they are being objective enforcers? Maybe, maybe not.

Once the dust settles on these issues, this bad news about partnering seems to really be good news. With partnering, the project is usually finished earlier with better results and with better compliance to our government rules and regulations. The only bad news is that it confuses the liability lawyers and the insurance people and triggers the public's paranoia about getting ripped off by big business. The taxpayers put our agencies in place to ensure compliance by inspection and en-forcement, which works fairly well. Now the enforcers can achieve the same end in a different way, but they are suspect to some interveners and may have to revert to inspecting, enforcing, and staying at arm's length.

However, overall we think that partnering is good news, especially for owners. Litigation on partnering jobs has all but disappeared. Liability issues aren't turning out to be as bad as was once thought. We still need to reallocate insurance premi-ums and costs, but we have always covered them one way or another anyway. It is probably just a matter of working out the details and getting on with the show. And the public is fortunately focusing more and more on results than process. We need a few more years under our belt to resolve these issues, but the outlook is certainly encouraging.

11

Benefits for the Architect Engineer

There are significant benefits that can accrue for the architect engineer who is working in a partnering arrangement.

One of the direct benefits of being a member of the partnering team is the understanding and knowledge of the client, its policies, procedures, and work processes.

Another direct benefit is being able to work with the other team members and share in their understanding and knowledge of the project and the construction process. With the open communication shared by the team and a clear and early definition of the project, he or she can focus technical knowledge on the areas of preliminary design, engineering, and constructability.

In the preliminary design, the architect engineer can contribute to a clear determination of design and engineering criteria for a project in the following areas:

- Feasibility
- Justification
- Background information
- Description of exiting and proposed facilities
- Process flow diagram
- Basic instrumentation requirements
- Material and energy balances
- Alternatives considered
- Physical property data
- Preliminary plot plan

- Equipment spec sheets
- Safety, environmental, and operational review
- Estimates
- Schedules
- Alternatives

Another area where the architect engineer can benefit from partnering is constructability. Constructability programs and reviews contribute to effective project execution by reviewing in advance the effect on construction of all project decisions. Constructability is important during the conceptual planning and preliminary engineering phases of a project because it alleviates the higher cost associated with changes or rework at later times in a project.

Areas where the constructability provides benefits to the architect engineer include the following:

- Establishing project goals and objectives. Design, engineering, procurement, and construction goals can be identified, established, and integrated in concert for the project.
- Early recognition of the client's specific requirements or methods such as specific safety requirements.
- The development of clear project specifications by having construction knowledge input into the design process at an early stage.
- Providing for the integration of design, engineering, and construction processes on the project to facilitate the schedule and minimize cost.
- The avoidance of project cost and schedule problems by having field construction expertise involved in the design process to consider alternatives involving construction methods, technology, techniques, materials, or equipment to be used in the project.
- Improving the understanding of design intent by construction.

Another direct benefit for the architect engineer is the ability to build and train a staff who then can provide depth and momentum in his or her organization for other projects that are awarded.

In some cases, the architect engineer may benefit by being placed in a position where he or she is made aware of future work in the client's strategic capital plans. This usually happens because of the high level of trust between the partnering members and the client.

In summary, the architect engineer benefits from project planning and constructability reviews by having a clear definition of the project and by the early determination of issues and problems that might occur during the execution of the project. In addition, being part of a team with open communication increases mutual understanding and knowledge of the project and of the important role each team member has on the project. The architect engineer as a stakeholder can bring value to the project and benefit from its successful completion.

12

Benefits for the Contractor

A contractor working in a partnering environment profits by overcoming some of the instances of wasted effort that normally occur under an adversarial working arrangement. With partnering and the team approach, some of the areas we have found where the contractor benefits are as follows:

- **Speedier issue resolution**—Within the partnering environment, we have found that many of the issues that would surface in non-partnering alliances have already been considered and resolved prior to construction commencing. For those issues that do arise, the openness and the group effort to reach its goal facilitate decision making and resolution. Issues are no longer considered adversarial.
- **Reduced rework**—We have found that the earlier involvement of the contractor in careful planning and constructability reviews reduces the amount of contractor rework on the projects.
- **More effective labor and equipment utilization**—Again, we have found that contractors have been able to obtain higher utilization of labor and equipment because of the increased planning and constructability reviews. Also, we have found that when a project has speedier issue resolution and reduced rework, these factors also contribute to more effective labor and equipment utilization.
- **Reduced claims**—We have found that there is a significant reduction in the filing of claims under the partnering approach. This is particularly true on the lump-sum contracts where claims are usually filed for changes in scope, or owner holdups. We believe that this is because of the openness, trust, goal focus, and the resulting planning and constructability reviews that mitigate late

owner and engineering changes. In addition, claims are reduced with better coordination of the procurement and construction schedules.

Another area where the contractor may benefit is the incentive-sharing formulae for the project. Incentive plans in partnering contracts are highly variable and client-specific. When the incentive criteria are achieved, the contractor can profit monetarily from his or her performance and from the overall project success.

13

Who Should Be Involved in Partnering?

Partnering's focus on extraordinary accomplishment involves a myriad of key players.

To pursue extraordinary accomplishment, the group must know what extraordinary accomplishment is and how to weigh alternatives against it. Defining the final result expected from the group is the beginning. In most cases defining the result is primarily a function of the owner. However, to best represent what they want and what will be in their best interests, owners may need fresh input from their key customers, suppliers, regulators, investors, community representatives, contractors, and union representatives. It is not always clear how well prepared an owner is to represent these outside views. The EPC contractor leading a partnering effort may be well served to request that one or more of these other players join the group, at least for early review and planning sessions. To the extent that throughout a project the partnering group will be seeking to enhance the project on the owner's behalf, it may also be prudent to keep some of these owner-related people in the group as associate members throughout the project, or at least invite them in to key meetings from time to time.

Various staffs from the owner's operations are also critical to determining how best to serve the owner's interests. Engineering, contracting, and financing representatives are critical to laying down requirements and limitations for the work. Input from operations and maintenance crews can be critical in discovering and working out breakthroughs and in heading off downstream problems. If the owner's procurement, safety, or quality assurance staffs will be central to the work, their participation will also be important. Some participation by labor representatives can also pay handsome dividends if labor problems or disputes are a threat to final results. Keeping the labor force informed of what is going on and engaging them in safety and bonus programs can be key. The EPC contractor must decide if, when, and how such labor representation and communication is warranted. If it is, it can be accomplished through existing partnering participants, unions, labor relations, or direct contact with labor representatives.

The owner's financial staff has two important roles to play in partnering. First, the EPC partnering group will be concerned about providing essential accounting and cash-flow information to the owner. Final accounting and tax records can be especially important to some owners, and if cash-flow management is an issue the group will need to be aware of what is and is not allowed. Second, the owner's financial staff will be crucial in helping the partnering group to build and use an economic model to assess various ideas and probable outcomes. Some suggestions are hard-dollar-based and easy to resolve one way or another. But some suggestions involve long-term operations, and environmental and marketing issues. Determining whether or not ideas are in the owner's best interests can involve fairly complex modeling of economic impacts and probabilities. Although the EPC contractor's staff will generally develop most of the needed models and provide most of the input, the owner's staff should be involved to review and enhance the analyses from the owner's side of the table. By engaging the owner's financial staff in the partnering group, the models and how they are used can be improved, and the owner's understanding of the EPC contractor's presentations can also be improved. Not having the owner's financial staff associated with the partnering group can create an adversarial relationship wherein the contractor presents a case and the financial staff seeks to disprove or discredit it. Working from the same side of the table on financial and financial analysis issues brings the same benefits to financial matters as it does to the basic EPC work.

Management reporting and public relations can be critical to major EPC projects. Engaging an owner representative to work with the partnering group to keep management reporting and public relations at the forefront can be a good move, especially if the project is large or complex. Such a participant can be an associate member of the partnering group. They do not need to be a full working member of the group, but they should attend all of the training and development workshops and most of the major partnering group meetings. Their role is to ensure that management and public information is adequate, appropriate, and timely. Included in this responsibility is liaison work with regulators, legislators, lobbyists, and interveners. To the extent that these relationships are critical to the project, then the participant should be a full member of the partnering group.

Occasionally the EPC partnering contractor will want to set up associate participation by industry gurus and key players associated with other facilities or projects. Their insights, ideas, regrets, and suggestions can provide important food for thought. Partnering seeks to accomplish the extraordinary, and sometimes good ideas and suggestions can come from those who are operating similar facilities or have recently built a similar facility. Consultants, vendors, and engineers who may have new materials, equipment, or methods may also be invited in from time to time to participate in brainstorming sessions. Beyond gleaning good ideas from these associates, time spent associating with these outsiders can be invigorating for the partnering team. Their recharged enthusiasm to think creatively can pay off later in their day-to-day work.

Partnering almost always involves significant measures for dispute avoidance and alternative dispute resolution. To the extent that dispute development can be turned into problem solving and corrective measures, those involved will likely be at least associate members of the partnering group. To the extent that legal representatives from each of the major views are willing, they can also be brought on board as associates to creatively head off and resolve problems before they become legal matters. At the genesis of partnering in the Corps were legal minds that switched gears from looking for and pursuing legal actions to looking for and resolving potential legal problems. From their early efforts it is clear that avoiding and resolving problems can be every bit as challenging and rewarding as taking opponents to the mat and winning. And it is far less costly. Avoiding and resolving problems requires creativity and problem solving that transcends standard legal practice, and that can be appealing to many.

CONSIDERING PROFESSIONAL FACILITATORS

Partnering takes on the challenge of working with all the different perspectives and resources that are critical to the outcome. One of those perspectives and views is awkward at best. It is partnering. Engineering is important, procurement is important, construction is important, and partnering is important. Partnering needs to be managed just as all the other aspects of the EPC project need to be managed. While partnering is taking care of everything else, who will take care of partnering? Partnering cannot be left to chance. Its own quality, timeliness, and productivity need to be tracked and managed. Partnering needs to be under the same management regimen as the other disciplines: planning, doing, checking, and adapting. However, in the frantic pursuit of extraordinary accomplishment, can any one player with their own role to play in the project also take on managing partnering? We think not.

We believe that one or more of the participants need to focus on how partnering is doing and what might need to be done to keep it strong and improve it. Such a participant would coordinate training and development, administer and process partnering evaluations and assessments, coordinate feedback sessions, ensure follow-up on action plans having to do with the partnering process, and be available to anyone inside or outside of partnering for information or counseling. Leaving partnering up to chance or making it a side job for an otherwise busy participant can be the downfall of a partnering group. The partnering process manager could be one of the associate participants, but probably the wisest choice is a professional facilitator experienced in partnering.

Understanding partnering is a challenge, getting others to understand it is a bigger challenge, and getting others to do it well is a monumental challenge. A successful facilitator must have an impressive mix of competency with partner training materials and exercises, meeting facilitation skills, interpersonal skills, brainstorming skills, focus group skills, analytical skills, and a wide range of insights that come from extensive EPC work. Quality circles, TQM, reengineering, and most of the other work-based management methods can do well by developing their coordinators or facilitators as they move forward. That is not the case in EPC partnering. In almost all cases the participants come from diverse organizations, have a wide range of motives and objectives, have not worked closely together before, and face a major project that is in a big hurry. The biggest risks and opportunities fall early in the project, so partnering has to be at its best right out of the gate. The partnering manager has an enormous responsibility to get the group on its feet and all running at the same target as quickly as possible. They can't be thinking about other responsibilities; they can't be learning about partnering. They have to be doing it and doing it well.

Having expert facilitation at the onset of partnering may well be half the battle. Done right and done quickly, partnering can achieve its full potential. Otherwise, partnering's total effectiveness can be cut to half or less. A good facilitator not only leverages the participants to maximum performance, but also spots and resolves performance pitfalls and inhibitors. Getting the partnering group to think and act as one is no small task. The most obvious and most used source for facilitators is professional consulting and service firms. Although not usually the case, excellent facilitators can also come from players in other successful projects, usually the EPC partnering lead. If the facilitator has prior or existing ties to any of the participants, their role can be awkward and even jeopardized. Laying such possible or perceived conflicts of interest out on the table early is important for establishing trust. If the group discovers "hidden" ties later in the process, trust in the facilitator can be irreparably damaged. We cannot overemphasize how important a good facilitator is to partnering's success.

CONSIDERING ARM'S LENGTH RELATIONSHIPS

Not all ideal participants can join the partnering group. For legal, liability, political, or regulatory reasons, some participants have to bow out who could otherwise be valuable contributors. At times this arm's length requirement makes sense; some checks and balances in our society are in our best interests. But at other times this seems ludicrous. In almost all cases their objectives, the objectives of those they serve, and the project's objectives could be best met with mutual cooperation and planning. However, our society does not always view these things the same way as the owner and EPC contractor.

With good persuasive arguments and careful public relations, some of these arm's-length relationships are changing. As we see more and more successes, perhaps we will be able to see more and more watchdogs moving around to the same side of the table as the partnering group. Joining forces to figure out what is best for everyone simply makes good sense. Until we can join forces, the partnering group simply does what it can to smooth out the relationship and ensure that it works as well as possible in producing the best possible results.

HOW MANY TO A GROUP?

Along with who should be involved, we should probably comment on how many should be involved. Facilitators will help design the partnering groups for best performance. They will consider how many possible participants there are, the complexity, magnitude, and pace of the work, their own capabilities, and the capabilities of the key participants. In most cases the partnering group will be one set of individuals numbering from 3 to 30. Most effective groups seem to have between 5 and 15 fully committed members. For the considerations listed, the larger groups may be divided into a parent group and one or more subsidiary groups with at least a third of the members of any subsidiary group also being members of the parent group. In rare cases only one member of a subsidiary group belongs to the parent group, but in most cases it is advisable to have at least two common members. When the division into subsidiary groups is a function or depth issue, all members of a subsidiary group may be members of the parent group. The division is done in this case to relieve a majority of participants from long in-depth discussions that are relevant to only a select few.

However, dividing the partnering group into subsidiary groups should only be done when the need to do so appears great and not doing so is likely to jeopardize best results. When in doubt, we believe the group should remain intact. Parts of the discussions and partnering work can always be done by subsets of the participants. Camaraderie, trust, and enthusiasm invariably suffer with each division of the group. Therefore, the drawbacks of dividing the group should be weighed very carefully against the benefits before dividing the group.

WHO TAKES TO PARTNERING, WHO DOES NOT?

It is hard to predict who will take to partnering and who will not. On the one hand, there are quiet but valuable people who blossom under the open partnering culture. Partnering thrives on and assertively seeks open and complete communication, and that draws the quiet and timid people out into the mainstream. Their ideas, concerns, and input are sought in a non-threatening setting, and they respond.

On the other hand, there are the outspoken, opinionated, hard-nosed types who initially have a really difficult time in a partnering setting. They too blossom under the partnering culture, albeit in a different way. They come to realize that they can be more effective than ever by taking a strong position in the partnering group and leveraging everyone's creativity and input to the hilt. Instead of just using their heads and the heads of a couple of close associates, they discover the power of using all the heads of the group. With competent facilitation, partnering can absorb just about anyone who can buy into a common goal and make it at least halfway to engaging in open two-way communication, and that includes the hard-nosed ones. Admittedly, the hard chargers seldom make it to angel status in the partnering group, but their strength and forcefulness are valuable and necessary components in the partnering equation. There are a few people who can't, or won't, make the transition to partnering, but with time and competent help most who give it an honest effort do make it.

The fastest transition to partnering is usually made by those who are already seeking open and honest working relationships and get a kick out of rallying with a group to take on a major challenge. They usually enjoy the ups and downs of creativity and taking chances, winning and losing along the way, and sharing with the group in the risks and rewards on both a short- and long-term basis. These are the people that are already cut out to get involved in partnering. Their learning curve is much shorter than what most of us face.

When we look at the best talent in our workforces, we generally find a couple of important characteristics that are detrimental to partnering. For example, successful talent is likely to be more competitive than cooperative, and it is likely to be more into enhancing itself than enhancing others. In most work settings talent that is competitive and self-serving wins out over the talent that is not. Talent tends to want to express itself, so with time it learns to be competitive and self-serving in order to survive and prosper; it is an evolutionary aspect of our culture and commerce systems.

It just so happens that when talented individuals are brought together and allowed to be cooperative and group-serving, they can be much more effective than talent that is brought together but is competitive and self-serving. To succeed, partnering needs to optimize the talent available to it. To do that, partnering must set up and enforce a system that (1) recognizes and rewards talent that cooperates and serves the group and (2) discourages talent that is competitive and self-serving. It must then work with the individuals to help them identify where their behavior is detrimental to extraordinary accomplishment and how they can transform the group to the cooperative group-serving behaviors that will best serve it's goals. Getting the maximum value from the best talent is critical to partnering's ability to rise above mere good performance, so this transformation of talent behaviors cannot be left to chance. The transition and management of the transition must be well designed and executed. At a minimum this means that detrimental behaviors must be recognized for what they are and dealt with in the normal course of working together. Properly designed partnering self-tests, surveys, and meeting guidelines are good first steps, and these are made immensely more effective when facilitated by professional help.

MORE ABOUT HARD CHARGERS AND DIFFICULT PEOPLE

In selecting who will participate in the partnering teams, it is common to find that the go-getters of the past are often placed low on the prospect list. They are considered too opinionated, too strong-willed, and not good team players. It is important in looking at these people to consider why their behavior has been as it has. We humans bring a wide range of coping skills to the table, and we bring them to the table for a variety of reasons. It is true that some ambitious persons would be more of a problem in a partnering group than they might be worth. But some go-getters are exactly what partnering needs.

One reason that some people become opinionated and make poor team players is that they have a burning desire to accomplish the extraordinary. In their experience they have found that bowing to group thinking and protocol has meant getting less done than just sorting out what needs doing and charging ahead. Being a good team player has meant giving up extraordinary accomplishment for middle-of-the-road performance. Their desire to accomplish as much as possible has driven them to take charge and take the heat for forcing their way toward the goals they believed could be accomplished. Not being a team player was the price for extraordinary accomplishment.

When introduced to partnering, these peak-performing go-getters may realize that they are in a group designed to accomplish the extraordinary and they can become invaluable team players. Those who knew them to be independent enterprisers are amazed at their transformation into enthusiastic team players. In truth, it is not the player who has transformed. They are still doing what they believe is best for accomplishing the extraordinary. What has transformed is not the go-getter, it is the surroundings. Associated with a group that seeks extraordinary accomplishment and has a low tolerance for detrimental policies, protocol, and motives, the go-getter enthusiastically joins the group. He or she who was formerly a team player on a team of one becomes a team player on a team of many. It is also interesting to note that such a person often has far less difficulty rising and adapting to the partnering challenge than do some of his or her counterparts who were good team players but were not as committed to extraordinary accomplishment.

When in doubt about whether or not to engage someone in partnering, it is usually a better policy to bring anyone on board who could be a solid contributor and then let partnering take its course. Most good players will adapt well to partnering, and those who do not will be routed out by the group. When any individual is detrimental to the group, the problem will eventually be raised and dealt with. Carrying detrimental participants is not a burden that partnering will endure. The pursuit of extraordinary accomplishment will take precedent over unreasonable tolerance of a participant's ego or shortcomings. Partnering has a heart, but it has its limits when it comes to any significant interference with accomplishing the extraordinary.

Partnering is obviously a group effort. However, partnering is not just following a plan. It takes a special level of trusting and straightforwardness that doesn't come easily to many people. It also takes a special willingness to question tradition and ask "Why?" and "Why not?" Not everyone in the group will buy in at the start, and not everyone has the staying power to see partnering over its first few hurdles. It can happen that the staying power for the group to rise above mere good teamwork may come from only one or two individuals. When this is the case, these people will end up putting nearly all their time into coaching and counseling until the others start catching on. Launching a spirit and a style is much more difficult than most newcomers realize.

A considerable part of succeeding at partnering involves a willingness to identify and face shortcomings and doubts. Raising such issues is usually not a wise move in most work settings. When faced with shortcomings, we end up looking the other way, pretending we didn't notice, and sidestepping the issue. But this has a detrimental effect on extraordinary accomplishment. Partnering participants need to recognize detrimental behaviors and get comfortable dealing with them head-on. In some ways partnering meetings are a little like Alcoholics Anonymous meetings where helping people recognize their problems and dealing with them are important functions of the group. Clearing out problems frees up an amazing amount of time, energy, and creativity. Whoever joins the group must successfully join in this straightforward approach to identifying, accepting, and resolving shortcomings in one's self and the group.

14

The Partnering Process

A number of times we have referred to partnering as a process, but suggesting that partnering is a process is misleading. When we talk about a partnering process, we find that listeners begin noting the steps in the process. They expect to do the steps and thus do partnering. Their thinking is that if it is a process and they follow the process, they then will succeed. Following the process we are going to present will set the stage for partnering, and it will enhance the likelihood that partnering will emerge and be successful. But partnering is not a process; it is a set of beliefs and behaviors that springs from the process.

If there is a core process to partnering, it is the standard EPC process. Starting from contracts and governing documents, the EPC contractor plans, coordinates, and controls the project. There are quality assurance processes, accounting processes, billing processes, permitting processes, scheduling processes, subcontracting processes, change order processes, approval processes, and so on. Partnering or not, the EPC work must be done and done right. Planning, doing, controlling, monitoring, checking, and reporting require step-by-step processes to ensure that everything is set to happen when it should happen in the way it is supposed to happen so that the final result emerges to everyone's satisfaction. There are even other processes to check the processes, and other processes to fix the processes when they fail.

These processes are essential to solid EPC work no matter what the work style is. However, through partnering these processes become less costly, less time consuming, and more effective. But partnering does not replace these essential processes. Lest anyone believe that partnering is a whole new way of executing EPC projects, it is not. It is a new way of managing and facilitating the EPC work, but it uses essentially all the processes normally required in EPC work.

Unfortunately we cannot just spring directly into partnering, so we have a process that will lead us to the edge of partnering. With hard work and a little luck, partnering will be born from the process, and the process will have been successful. However, partnering is difficult to maintain. There are about as many forces working against it as for it, so we will talk about another process that helps us keep partnering on its feet. And again, that process is not partnering. Partnering is a work style that needs processes to help it along, but it is not a process.

Now that we have emphasized that partnering is not a process, we can begin our discussion of the "partnering process." No matter what we say here, most people are going to refer to partnering as a process, so here we go.

Those who are already deeply engaged in partnering will have gone to much more depth than we are going to cover here. Those who have contracted for, or will contract for, professional partnering facilitators will also have access to their detailed workplans and materials. Our intent here is to provide a primer for those who are at the fringes of partnering or are contemplating jumping in. The following steps outline the essential phases and considerations and provide a sampling of what is involved. If you are involved in partnering and discover that your program is perhaps not up to par, that should spur you to research and resolve the situation. If your partnering is not up to par, then your results are in jeopardy.

TO PARTNER OR NOT TO PARTNER

Owners and EPC contractors who are experienced with partnering can usually sort out the pros and cons of partnering on a given project and simply get on with it one way or another. But those with limited or no experience can face quite a dilemma. As a general rule, if the project will take six months or more to complete and it involves at least two or three million dollars, then establishing a partnering arrangement is probably worth serious consideration. Owners and EPC contractors with partnering experience that can be applied directly to the project at hand will typically use it with even smaller projects, but first-timers need a little breathing room to bring partnering on board and a substantial enough project to make partnering worthwhile. These guidelines are offered because people often ask questions about them. In reality, doing or not doing partnering depends most on who is involved and who makes the decisions.

Certainly there are other considerations. If the work is one in a series of cookie-cutter projects that are cut and dry, then opportunities to save the day or identify significant breakthroughs are fewer. In these cases the cutoff for partnering might be increased to projects of a year in duration and four or five million dollars in cost.

When in doubt about using partnering, owners and EPC contractors might want to consider a phased approach. Commit to a one-month trial period with a moderate partnering effort overlaid on standard EPC contracting. The partnering will slow standard start-up to some extent, but the cost and delay are rarely a complete waste, and if the program takes off the benefits will be well worth the risk. See how it goes. If it goes well, continue the partnering overlay or expand it. If partnering does not appear to be going well, back off and revert to your standard EPC project practices.

Launching a full-blown partnering effort is almost always best, but half an approach with at least some of the benefits is better than none, and everyone can learn from the experience. However, a caution about reduced approaches and overlays. There is a tendency to use less than the best facilitators for the reduced effort. Doing half an approach with an underpowered facilitator is a serious mistake. If you are considering a reduced overlay approach to test the waters, use a well-experienced and successful facilitator for your effort. Anything less almost ensures failure.

Pick the People

The first and most important thing to do is select and bring on board the partnering facilitator. After reviewing the project and potential participants, the facilitator can recommend who they believe should be included in the program, whether or not there should be more than one group, what steps should be taken before a launch, and how best to launch the effort. The owner and EPC contractor can jointly consider the facilitator's recommendations and work out any details necessary to identify and commit who will participate.

Prepare the People

In some cases all participants will show up at the launch with no preparation. All training and development will start during the launch workshop. In other cases some or all of the participants will be given reading materials with an expectation that they will actually read them before the meeting starts. In some cases some or all of the participants will be asked to complete workbooks that may or may not need to be sent in ahead of time to the facilitator. These workbooks usually have to do with understandings about the project goals, risks, and key issues or critical success factors. It is also fairly common to have participants fill out personality questionnaires and send them to the facilitator before the first meeting. The questionnaire responses are tabulated by the facilitator and used during the first meeting to get into interpersonal working relationship issues and loosen the group up. If any homework is due from the participants before the meeting, the facilitator will certainly have someone tracking the requests and returns to ensure that each participant responds appropriately. Not having a homework monitor will almost always result in a low response.

Regardless of whether or not participants have homework to do before the meeting, the facilitator will work with the owner and EPC contractor to design appropriate press releases, newsletter items, road shows, executive briefings, and internal announcements to introduce the partnering effort to the organizations involved and pave the way for the effort. If participants were not contacted during the member selection process, they and their supervisors should receive written and personal contacts regarding their selection and what will be expected of them. Before sending out homework requests, participants should receive memorandums describing what is coming, why it is important, and what is expected.

The Launch Workshop

Almost every experienced facilitator has a different approach to the launch workshop, and how they engage the participants in the process can be quite different. However, the goals are essentially the same in all cases, so generally the workshops are more alike than different. Some launch workshops are only a day long followed by one or two boosters over the next few days, but most initial workshops cover two or three days. To accommodate differing complexities, sizes of projects, and participant backgrounds, facilitators may scale their launch materials down to less than a day or up to a full week.

A full-scale launch involves understanding partnering and how to proceed. It also involves a considerable amount of time understanding what is involved in the project and what it will take to succeed. And overshadowing all the workshop activities is team and trust building for the group.

Workshop Plan Adults like to know what they have gotten into and what is coming down the road. They like to know what the rules are so they don't embarrass themselves, and they want to know what provisions have been made for special needs, like rest rooms, phones, and messages. Reviewing the plan of the day is always a good starting point as is introducing the facilitators and any observers or nonparticipants who are in the session.

Participant Introductions In one way or another the participants need to get to know who else is in the room, who they represent, what their expected role will be, and what experience they bring to the group. Introductions are often the first item on the agenda, but some facilitators believe that it is a downer to start with a long series of introductions. They prefer to indicate that introductions will come later and move directly into more interesting or lively material. Until introductions are made, the facilitator calls on participants by name and asks for their names when they first speak. Full introductions are then usually covered after the first break.

The Challenge A typical starting point for the initial workshop is an exercise to ensure that everyone understands and is thinking about the same project. This also provides an ice breaker to get people talking about and debating some of the key points. The facilitator usually has the group offer bullet points to describe and bound the project. Once the project seems to be fairly well described, the facilitator asks for bullet points for a second chart on the biggest issues the project will face, followed by similar charts on the forces that are for the project and the forces that are working against the project. The challenge definition session usually closes by having the group make a first pass at a mission statement for the project.

Definition of Partnering   When the big picture is fairly clear and introductions have been made, the next hurdle is to gain a common understanding of what partnering is and what it is not. The facilitator either asks experienced people to offer their views or asks for input from the whole group. It is usually preferable for the facilitator to get a good sampling from both groups to avoid symbolically raising the stature of experienced people over inexperienced. Having a few "wrong" answers also helps the group learn how to confront one another with differences of opinion and yet still finish amicably. Although the facilitator will prod the group to untangle obvious discrepancies, these initial lists should not be perfect. Circling back from time to time to improve the definition of partnering is an important lesson for the group.

Personality Profiles   Most facilitators use standard personality questionnaires or normalized team or leadership surveys to help their groups understand their personality mix and help each individual understand what interpersonal dynamics will likely be most important to them. Such exercises also introduce the participants to openly discussing and dealing with somewhat sensitive personal issues, an essential learning for partnering participants. Some use instruments that are administered and scored before the launch; others administer and score instruments during the session.

Except for the most sophisticated clients who use personality training on a regular basis, instruments that are used during the session are usually scaled-down versions that can be completed and tallied within an hour and effectively used to cover group and interpersonal dynamics within two or three hours. For different circumstances facilitators will use shorter or longer versions. However, the intent in the launch workshop is not to overwhelm the participants with deep and complicated self-analysis and development challenges. In the launch session we are looking to address basic interpersonal dynamics and nudge the participants onto a more personal playing field with their teammates. Discussing and sharing sensitive information is a big first step in introducing the group to open and straightforward communication. More sophisticated interpersonal skills can be developed in future partnering meetings.

Philosophies and Values   Along the way partnering will face many situations that will require choosing between options. Some of those choices will involve hard dollars, and the choice will be easy. But some of those choices will have soft-dollar and value factors that are much less easy to call. To help the group begin to understand these important trade-offs, the facilitator will ask for the group to list all the major views on the project (including customers, suppliers, trades, regulators, etc.) and then ask the group to provide what they think is most important to each view. Once the lists are developed, the facilitator asks the group which views and values are most important, and which will have to take a backseat. Out of the discussion will also come basic regulatory compliance as well as legal and ethical bounds that must be honored. The intent of this exercise is not to actually develop an ironclad model for trading off one view or value against another. The intent is to help the group realize the breadth and complexity of considerations that will likely be involved in some of their future decisions. The exercise also provides a foundation for understanding more about each other's views and where significant differences of opinion may exist. Recognizing and dealing constructively with differences in the group is an important part of each exercise.

Critical Success Factors   Based on the prior materials and discussions, the group is usually asked at this point to sort out those few things that will seriously jeopardize success if they are not done exceptionally well. There are two types of critical success factors: first, those factors that improve performance the better they are done, like open communication and coordination, and second, those hygiene items that if not done to a required level will bring the project down (but doing them better than they need to be done does little for the project) like recordkeeping or review meetings.

The facilitator often has to lead the group to think along two dimensions. The first is getting the job done to expectations and specifications, which is the basic EPC work. The second is finding and incorporating breakthroughs that greatly enhance the outcome, which is more in line with partnering's goals. The group will usually note important differences between the critical success factors for each dimension.

Partnering Mechanics   The group is next challenged to understand how partnering will work, what it will take to get it working, what it will take to keep it working, and what process can be put in place to ensure that partnering rises to the challenge. Most facilitators use their own customized model for partnering, but the basics are quite similar.

In its most basic form partnering requires that the participants regularly engage in open and straightforward communication about what will constitute the best outcome under the circumstances, and what, if anything, can be done to improve final results. Riding on the back of standard EPC processes, partnering has an unending agenda to find the next best action which will improve the outcome. High on the agenda is finding creative breakthrough opportunities to leap forward; immediately behind that are creative breakthroughs in improving schedule and cost performance; and immediately behind that is heading off potential problems before they materialize. The partnering group is dedicated to working as one and optimizing everyone's input to improve on a baseline of EPC excellence. To accomplish the extraordinary, partnering uses a continually repeating two-cycle process aimed at accomplishing the extraordinary: (1) find the next best action to improve the final result, and (2) do it the best way possible.

The group is led through an exercise to determine how this will work in their day-to-day work. They will usually work together to create ideal agendas for scheduled meetings and ad hoc meetings. They will typically list the most common questions they will ask each other to keep all the important issues and questions at the forefront and to keep each other on their toes and thinking. Toward the end of this exercise they should also develop a questionnaire that they can fill out from time to time to see how well their two-cycle process is working. In the launch workshop it is usually better to go for participation and buy-in and leave corrections and questionaire cleanup for a later session.

The group can best address how to get partnering working by listing their group and individual strengths and shortcomings relative to their responsibilities to the project and to partnering. To get the project and partnering going obviously requires leveraging their strengths and resolving their shortcomings. At this point the facilitator can usually schedule a number of specialized workshops for selected individuals as well as several private counseling sessions and outside training programs. Getting comfortable with open communication, going around organizational protocol, raising skills in constructive confrontation, leading brainstorming sessions, leading focus groups, and working out interpersonal issues are usually hot topics for early training and development. The conclusion of this session requires that the group and individual participants sign up for actionable items on an agreed timetable. Recall that partnering is a work style and a set of behaviors that work best by going straight to the best results under the circumstances. Developing the work style and behavior sets as quickly as possible is critical to launching partnering.

As to what it will take to keep partnering working, the facilitator will generally work with the group to introduce a set of surveys and follow-up meeting agendas to regularly test partnering and the various forces acting for and against it. The purpose of the surveys and meeting cycles is not to maximize partnering itself, but to maximize the group's effectiveness in doing the best it can. There is no training for the sake of training just because of a survey response rating or a weakness in performance. Training and developing for partnering is only done if it is believed that it is necessary to achieve extraordinary accomplishment.

Given that the EPC group is almost always overworked and not inclined to go off-line unless absolutely necessary, the facilitator must be skilled in clarifying how different training modules can produce results for the group. If the group becomes disillusioned because they are caught in a business-as-usual slump and don't have time for real partnering, the training modules can also serve to recharge their partnering batteries and rekindle the partnering spirit. From time to time the partnering leaders may have to mandate survey and feedback cycles and training, but for the most part the group and individuals in the group should request the help because they believe they will produce better results for having received the training. Partnering will be kept working by having the facilitator continually working in and around the group and focusing on partnering's needs to accomplish the extraordinary.

When all this is in place, partnering should come up to speed and rise to the challenge. However, central to almost everything so far has been the facilitator. The adequacy of the facilitator and training will be a part of the regular survey questionnaires. If the group is inexperienced, they may not be able to tell whether or not their facilitator is up to the challenge. If the partnering group feels it needs help in evaluating the facilitator, it should arrange for another experienced facilitator to serve as a partnering auditor to review their progress and the project's facilitator early in the process.

When it is expected that an auditor will be used, he or she should be an observer during the launch workshop. Observing the workshop will allow the auditor an opportunity to offer suggestions to the facilitator during breaks. The suggestions can help the facilitator be more effective later in the workshop or the project and can thereby help offset the costs of the auditor. Observing the workshop will also provide the auditor with a basis for performing future audits. After the launch workshop, the auditor reviews the project and submits a report to the group, which will consider the audit report and responses from the facilitator and take whatever action, if any, seems appropriate. The auditor may try to unseat the facilitator. It will be up to the partnering group to sort out the situation in the interest of best results for the project.

Survey and Review Cycles

The facilitator is responsible for setting up and maintaining regular survey and review cycles. Without a disciplined and focused effort to continually test, tune, and adapt partnering, it will tend to lose ground against day-to-day pressures. Individuals will begin dragging their feet, then not show up for meetings. Meetings will be delayed. Then the group will fall apart and revert back to standard EPC practices just to get the job done. Partnering can pay off, but it can also fall apart. The facilitator is charged with providing the commitment and energy to keep partnering alive and well and doing what it is supposed to do. If the group comes to a point where it is convinced that the project will be best served by dropping partnering, then it should be dropped. But it should not be dropped just because it is having a rough time. The regular survey and review cycles serve to rejuvenate the group and allow it an opportunity to continually rise to its best level.

The initial survey forms and meetings will be as much ice breakers and training experiences as they are useful. The participants need a cycle or two to understand the meaning of the survey questions and how to apply the response categories. As the group comes up to speed on the shorter and simpler forms, the facilitator will change, add, and delete items to continually improve their value to the group. When an area of partnering is doing well, one summary question will usually suffice. When the area is showing signs of trouble, several questions may be needed to sort out symptoms from causes and get to what is needed to correct the situation.

The facilitator collects the questionnaires and to the extent agreed by the group protects individual responses from being revealed or deduced. One section of the survey addresses the individuals' perceptions of themselves and how they believe they are doing and where they may need help. Another section asks for opinions on how each of the other participants are doing and where they may need help. Another section asks for ratings of how the group is functioning, and a final section asks for views on what outside forces are working for and against partnering, including partnering's relations with key executives and other groups. Although most facilitators will start with a basic set of off-the-shelf questions and scales, these are usually customized from issue to issue to best meet the needs of the group in identifying what is needed to move forward.

The review meetings are used to go over the tallied and summarized survey responses and to work through the agendas first designed during the launch workshop. The meetings basically address what has happened, what was learned, what is next, who needs help, who can give help, how to best proceed, and what opportunities there are for project breakthroughs.

In-Services

At the beginning of a project facilitators generally provide a partnering handbook and offer a string of workshops and training sessions to address the most common training and development needs of partnering groups. To the extent that the series meets the partnering group's needs (and not the facilitator's needs), they should be followed. At some point the program designed at the beginning of the project will not be the best course of action and from then on the facilitator is responsible for offering the training and development that best meets the needs of the group. Although some of the training and development can be offered as a stand-alone workshop, at some point the bulk of it is offered in half-hour, hour, or two-hour chunks as adjuncts to other partnering meetings. These in-services are more flexible and less disruptive than classroom sessions. They are also more conducive to dealing with real situations, which makes the learning more meaningful and the training more directly beneficial to the project.

The facilitator will offer a string of interpersonal relationship building sessions as well as modules on most of the popular management tools such as quality circles, MBO (management by objectives), TQM, and reengineering. Partnering uses whatever works best under the circumstances, so some level of training on the most popular and useful tools is necessary for the group to be as effective as it can be.

Dealing with Participants Who Do Not Measure Up

Working past traditions and old behaviors to reach a new level of open and straightforward communication demands a lot from participants. To the extent that someone is not quite rising to the challenge but is not detrimental to the group, perhaps their participation can be continued. It depends on how much they are affecting the final results relative to what a replacement could do for the project. To the extent that someone is not rising to the challenge and is detrimental to the group, they should be replaced if at all possible. The genius and power of partnering come from the participants' extraordinary values and behaviors. For whatever reason, if a participant is not measuring up, partnering's single-minded focus on accomplishing the best under the circumstances means that business is business and the participant issue must be resolved. If there are no other choices and the participant is necessary to the group, then that is the best possible option for the best possible result, and the group simply has to get on with it. In trying to get the most done through organizations and positions it is working with, the partnering group will occasionally debate how certain people outside the partnering group should not be in the jobs they have and what can be done about it. Sometimes that debate is appropriate regarding their own members.

To salvage participants who need an extraordinary amount of help, evening and weekend sessions can be set up along the lines of support groups. The facilitator may get paid for these hours or may take compensatory time, but the attenders will likely be required to attend on their own time. When policies frown on such a volunteer off-duty program, the participants may need to be placed on probation, and volunteering to attend the sessions can then be a part of their recovery program. In any case, support groups can be very effective in helping interested participants get on track.

A Runthrough of Thoughts and Second Thoughts

As we have worked with partnering teams and engaged in roundtables and discussions, many topics have been played out. Following are notes from a few of those sessions.

Charters   A charter of philosophies and principles is relatively easy to draft and sign. Up to that point it is mostly just good talk and a dose of hype. But what happens when the group starts to walk the talk?

It is amazing how many groups come up with pretty good charters but don't seem to get very far. Actually, I guess it isn't too amazing when you take a look at their overall process. First, charters from these groups are often not so much from the heart as they are from the head and reflect what "should" be said. A fair amount of the best wording probably also comes from copies of other groups' charters. The words make sense, and the goals seem noble. But they are someone else's words and someone else's goals.

The final step in drafting a charter (just before signing it) should be to convert each key word and phrase into a survey question. Just before signing the charter, every participant needs to assure themselves that they understand what the questions and charter are asking for, answer the questions as they know the answers to be at that time, and make specific notes on what it will take for them to overcome any shortfall between where they are and where they need to be.

Having converted the charter to a test and having taken the test for the first time, the group is ready to sign the charter and move forward. Incidentally, one of the group's first steps in moving forward should be to schedule when they will have a group session to reread the charter and retake the test. The fundamentals of partnering require professional management: perpetual cycles of planning, doing, checking, and adapting focused specifically on partnering's effectiveness and success.

Partnering is essentially self-sustaining and self-improving, but it is still a system of people. If you fail to launch a management system right from the start, you seriously jeopardize your partnering effort.

Getting Up to Speed   Partnering does not happen just because we want it to happen and we assign good people to it. Partnering groups must come up to speed on partnering just as any professional team must come up to speed on what it is about and what it must do.

Independent reading and study is a good first step, especially if the group will not be having a general introductory course and will be going straight into a partnering workshop as a slated or potential participant. With a general overview of partnering behind them and perhaps a video or two, the group participants typically meet in a facilitated formation workshop. Depending on the facilitator, the group will sort out roughly what partnering is and is not, and will move rather quickly into a first pass at statements of vision, mission, philosophy, values, and principles. By tackling all the major issues at the beginning, the group gains a fairly complete foundation to work from for the rest of the workshop. The beginning sessions are like the tip of the iceberg. What follows is several cycles of plowing through the same issues, but each time at a deeper and more meaningful level.

Interpersonal Skills   Interpersonal skills are the workhorse of partnering, but the facilitators will generally not deal with interpersonal subjects until the formulation session has been underway for a couple of hours. In most formation sessions most facilitators will also use a high-level personality test and exercise to help loosen the group up, introduce interpersonal issues, and begin building the basis for straightforward communication and trust.

With the first couple rounds of issues under their belts and some level of personality and interpersonal work behind them, most well-facilitated groups are already demonstrating the beginnings of partnering. They are beginning to raise issues that they would normally ignore (or even deny), and they are beginning to be sensitive to effective versus ineffective ways to deal with each other. The beginnings are certainly a little rough, and most everyone stumbles several times. But when the stumbles are looked on as just stumbles on the way to success, and not failures, the group's openness and forthrightness begin to blossom. Raising issues that would otherwise not be raised and dealing with them with the sole purpose to get beyond them to the business at hand is a stimulating experience, especially for the old hands of the industry. Inhibiting values and behaviors begin to quickly fade as the group sharpens its focus on results. As the facilitators ease the group through a series of exercises, the group gets better and better at spotting and resolving barriers to desired outcomes. Individual by individual the lights start coming on, and with every light that comes on the group's power to work together grows.

From Team Building to Partnering   A quick burst of enthusiasm and excitement typically emerges during most motivational workshops. Those who have been through team-building formation sessions before may well ask how this is different from standard team building. For the first couple of sessions there will be little difference between standard team building and partnering formation. But certain differences will become obvious as the facilitators take the group through exercises to specifically address cases where changes to policies, traditions, and protocol have yielded significant results. And then the group identifies where they may have opportunities to work through similar issues for significant breakthroughs. And the group will start brainstorming more and more for technological and methods breakthroughs. Whereas team building usually fires the team up to do a great job, partnering fires the group up to accomplish the extraordinary. The two challenges are much more alike than different, but they are different. Taking the partnering group up through great teaming to partnering as quickly as possible is the primary task of the partnering facilitators.

By the end of formation, standard team-building groups usually have a list of winning practices focused on project execution excellence. They are fired up and signed up to go out and do a great job. By the end of partnering formation, the partnering groups have a list of winning practices focused on accomplishing the extraordinary. For a partnering group, doing a great job is not the issue, contracts are not the issue, organizational responsibilities are not the issue, great teamwork is not the issue, and meeting job specifications is not the issue. The only issue is extraordinary accomplishment; absolutely everything else is open for review. The partnering training makes clear that the work will get done because that is necessary to extraordinary accomplishment, and what will be done will be ethical, legal, and honorable. But the group's individuals effectively shed their organizational loyalty and allegiances and realign themselves with extraordinary accomplishment, lock, stock, and barrel. In a sense partnering is great team building that rises to a new and higher level.

For those who appreciate the Maslow motivation hierarchy model, one way of looking at the difference between team building excellence versus partnering is that team building is at the esteem level where work itself along with recognition for a job well done are the motivators. Partnering is at the self-actualization level where achievement of the extraordinary is more motivating than recognition. It is not that partnering is better than standard teaming building, it is simply that partnering is the next step up the ladder of group effectiveness.

Training   As with any team building, skill building on hypothetical material is minimally effective. Putting hours and hours into hypothetical cases and simulations and then expecting a good transference to the real world just does not work well. For partnering to work well and meet its unique challenges, it needs to move as quickly as possible from theory and examples to real issues. Very important things happen when the group tackles real issues with real people's names associated to them and works through how they might deal with these real people. First, application is where learning really comes from, not studying. When the group studies a little and applies it, it learns a lot. Second, by working on real situations the training focuses on what is important and meaningful. Training on real issues and situations yields training for the sake of accomplishing something, compared to training for the sake of training. Third, in the course of training on real issues the group will invariably ask a lot of "what abouts." "What abouts" are often the agenda for the next training session. Fourth, training on real issues versus cases and simulations can often lead to real solutions, which significantly energize the group and help to offset training costs. Fifth, in the course of wrestling with real issues a number of unresolved issues will surface that can be noted and worked on after the training session. Each participant should be able to leave the session with a number of meaningful to-do's.

Training according to an agenda is easy and it ensures that you cover a list of known important topics. However, it is not necessarily the best way to train a group that is facing immediate important work. By letting real issues lead the group from training need to training need, the material is more meaningful, more relevant, and more learnable. A good facilitator will maneuver the group into discovering the really important lessons that need to be learned, those biggies on the original list. They will also weave together technical points, process points, interpersonal points, and philosophical dilemmas, and so on. We are redefining training for partnering to be more coaching than training. A few training modules can be effective here and there when the time is right, but hours and hours of programmed training are generally not as effective for the partnering challenge as are guided coaching and development.

How can such free-flowing training possibly be better than a well-structured program that cranks out high levels of learning? The answer is surprisingly simple. What the partnering group needs to learn as much as anything else is the discovering of issues and problems, discussing them, and dealing with them. Stumbling from one bump to the next and working out understandings and solutions on the run are important pieces of what partnering is about. Identifying issues, understanding them, and working through them are key. Getting halfway through one lesson and realizing that the group must backtrack for some other lessons before moving ahead is reality. It is an awareness and a skill that must be developed.

Juggling a half-dozen half-answered questions and working through confusion and panic to eventually reach an answer are reality. It isn't really confusion or panic or a mess, it is progress. It is a kind of progress that the group needs to get comfortable with and get good at. The little bumps and confusion and panic in the training sessions are nothing compared to what the group will face on the project. The training sessions are a first run at important lessons in partnering reality. So our suggestion is to bring in trainers who can roll with the group in a much more dynamic training session than what most trainers offer. Should you have a checklist of what should be covered in the session? Certainly. Should the participants end up with a handbook of principles and skill guides? Certainly. But let most of the training emerge from a realistic process of discovering needs and resolving them. Using a free-flowing training approach and real project-specific situations will yield a much better understanding and ability to apply the understanding. Usable solutions will likely come out of the training, and a multitude of group interpersonal dynamics (which will come up later anyway) will pop out during training. When they emerge, the interpersonal issues can be worked on in the session before they have to be dealt with in the field.

Executives on the Perimeter of Partnering   Executives that are not directly involved in the partnering group are sometimes put in a particularly difficult position. They are asked to believe in the value of partnering and support it, and if partnering is going well and everyone is thrilled with what is happening, there is no problem. But from time to time in order to accomplish the extraordinary, the partnering group works around the establishment or challenges it head-on. Executives at the borders of partnering can find themselves facing a difficult choice: siding with their organizations who want partnering to toe the line, or siding with partnering members who want to change policy, practices, and protocol to pave the way for extraordinary accomplishment.

When partnering can present a cut-and-dry case that is clearly good for business, the choice is easy. When partnering's case is more theoretical than hard-dollar, the choice is much tougher. We all understand that partnering will take a run at several breakthroughs and will only succeed with a few. We all understand that the magnitude of the few breakthroughs more than compensates for the false and failed runs. However, each individual run at a breakthrough is either a success or it is not.

When the run is not a success, the executive that sides with partnering against his or her own organization may be right to support partnering for the overall good, but it is a tough line to sell to his or her own troops. This is where partnering must rely on those executives who are working directly with the partnering group to chip in, mediate, and coach their fellow executives on how to play ball and cope with the consequences. The partnering group should be quick to acknowledge when their breakthrough runs don't work out as hoped and use the right channels to have those who backed them in their attempt appropriately acknowledged, thanked, and supported in defending their role.

Executives can find themselves in very uncomfortable situations when they are not directly involved in the partnering group but from time to time find themselves working with the group on major issues. This is particularly true for those executives who supervise members of the partnering group. The supervising executives invariably have to publicly support partnering even if they aren't quite sure it's all that it's cracked up to be. These supervising executives in one way or another have usually bragged about "their" partnering group and how the keys to partnering are openness, trust, and a willingness to face anything in the best interests of accomplishing the extraordinary. After having publicly and socially said the right words, the supervising executive often finds that those words come back to haunt him or her. Members of the team show up at their door and do just what the executive has been talking about. They are open, trusting, and willing to face anything in the best interests of accomplishing the extraordinary. And at that moment the barrier hindering extraordinary accomplishment happens to be something that needs the executive's decision and support. The partnering group cuts to the heart of the matter and asks the executive to take a stand to support a breakthrough. It may be breaking protocol, waiving a long-established but silly policy, or breaking new ground in empowering workers to make a series of decisions and do what they think is best. The point is that the executive is expected to be a good partnering player, but that's not necessarily a role they are at all comfortable with, and not a role they may even be able to accept no matter how many times they have said the right words.

When executives not engaged directly in partnering are being brought in on major issues and decisions, the key to success generally lies with the partnering group. At a minimum they should involve other executives that are involved in partnering to coach the non-partnering executive through the issue. If the issue is important enough or the non-partnering executive difficult enough, the partnering group should design a special workshop around the issue. They should use their facilitator and a special subset of partnering and non-partnering people to openly work through the issue. With adequate preparation, the facilitator should be able to get most of the hidden agendas and concerns out on the table and help the group work their way through them. Partnering groups can lose their sensitivity to how threatening the partnering style can be to outsiders who have a lot at stake.

Partnering's Obligation to Support Management   Some partnering groups complain that management doesn't give them enough support. In truth, these groups are slipping off the partnering track. It is partnering's responsibility to assist management and gain their support in whatever way best supports extraordinary accomplishment. Partnering doesn't have time to blame anyone or anything. The focus should be only on extraordinary accomplishment and what can best serve the most good. If the group needs more support from management, it simply searches for ways to get what it needs to move on. If they can improve their approach to management, that is the issue. If they can't come up with a more successful approach, then management support is as good as it is going to be and everyone gets on with the task at hand. In this case, if the partnering group loses its sensitivity to those who are not engaged in partnering and mishandles any dealings with them, then the loss of that sensitivity is what is interfering with extraordinary accomplishment. The solution is to rebuild the sensitivity and develop a better approach. Whether management is or is not supporting the right issues in the right way is not really the problem. The problem is figuring out how to best gain the support of management.

Scoreboards for Partnering   The genius of partnering lies primarily in its single-minded focus on accomplishing the extraordinary. Sports teams can focus on a scoreboard, and extraordinary accomplishment boils down neatly to which team has the highest score. The scoreboard for EPC partnering is obviously much more complicated; it tends to be more like several scoreboards all hooked together. The headings on the scoreboards are things like safety, compliance, schedule, quality, total installed cost, margins, and life-cycle economics. Scoring on any one score-board tends to affect the scores on other scoreboards. For example, improving safety usually requires a tradeoff against schedule and cost scores. Some of the scoreboards have two or more numbers on them. For example, life-cycle economics has a score for the owner and a score for each of the contractors involved in the project. In some cases all of the scoreboards can be rolled into a sophisticated algorithm and one score can be developed. But such algorithms are few and far between, and are usually too suspect to be effective.

Partnering's pursuit of extraordinary accomplishment is not a matter of achieving a single high score. You can't open the newspaper one morning to the project section and know exactly where the project ranks or how it came out in the final standings. Winning at partnering is more like day-to-day competing in the Olympics where you face several judges and more than one score. The team of judges may evaluate you first on content, then on form. Although every attempt is made to have a logical scoring system, some judge's scores seem to stand out from the others. Sometimes their scores are higher, sometimes lower. They seem to have a different set of criteria for judging than the other judges. We can guess at what their thinking might be and what is motivating them, but we will never know for sure. In the end, after days and days of performing and trying to please as many different judges as possible, the scores are in and a final standing is posted. Some achieve extraordi-nary status and most do not. The scoring for partnering is similar.

In the beginning the partnering group visualizes the final performance and how it thinks all the various judges will score the final project. The group does its best to understand how performance against any one score will affect the other scores. It tries to second-guess the various judges and what will best influence their scores. The partnering group builds a shared vision of what it will take to accomplish the extraordinary and how the final score will come together. And then the group launches its effort. Testing new ideas and evaluating whether or not to make a run at a breakthrough is difficult. The group must decide how the effort is likely to affect all of the various scores and judges. Some effects will be straightforward and easy to predict; some will be little more than an untested concept based on high hopes and determination. And like a venture capitalist's portfolio, the group needs to take on board several breakthrough endeavors to best ensure that the big hits will outweigh the misses. Aiming for extraordinary performance is clearly more of an art than a science.

At the beginning of a project the partnering group works through what it believes the various scoreboards will be and how they will work. They rough out how the various scores will complement each other and how they will detract from each other. The group considers the many judges of the final project and works hard at understanding each judge's view and how to earn the best score from them. At some point most groups are overwhelmed by what it will take to please everyone. But sorting that out is what it takes to build a sense of what the ultimate measure of success must be.

Owners must play a key role in defining and using the scoreboards for the proj-ect. Understanding how the owner views and will value the many tradeoffs in EPC work is critical to being able to choose how best to proceed with a project. The partnering team must be able to work closely with the owners to build and apply the scoreboards. Some of the scoreboards depend on hard dollars, so what they are and how to use them are fairly straightforward. However, evaluating breakthroughs and approaches also involves a tremendous amount of soft-dollar thinking. For example, safety is important, but should safety training be increased by 100 hours per quarter or not? On the one hand there is the EPC team's budget and hours lost to training versus minimum standards; on the other hand there is the owner's reputation and image regarding safety. If the labor market is under a lot of pressure and the ratio of green hands is up, will the owner share in the increased training costs, or do they want the EPC team to hold to the budget or absorb the cost?

It is how the owner scores the final project that matters most. Performance against all of the various scoreboards affects the owners in some way. If safety is poor, costs, schedule, and litigation work against the project. If environmental compliance is lacking, the project suffers. If contractor profits are too high, owner image and profits suffer. If contractor profits are too low, quality and schedule are likely to suffer. At the end of the day, it will be the owner's final opinion of the project that will determine if the results are substandard, ordinary, or extraordinary. Short of ethical and legal boundaries, and short of being grossly unfair to suppliers and contractors, the challenge of the partnering group is to maximize the owner's score of the project. It is not easy, it is a complicated mix of scores by a somewhat mysterious group of judges, but the focus for the ultimate score is on what is best for the owner.

Going Beyond Just Execution   Because what is best for the owner is the ultimate measure, partnering looks to go beyond mere excellent project execution. If the specifications can be changed up or down to enhance the owner's long-term satisfaction, then changing the specifications is a part of extraordinary accomplishment. If changing the design or planned execution could benefit the owner, then rehashing the plans is the issue, not just carrying out the contract. Because the goal is to accomplish what is in the best interests of the owner, the project plan and the EPC contract may need improving at the same time as the project is being built.

Partnering that looks for breakthroughs only in execution is doing only half its job. The other half of the group's job is looking for breakthroughs that go beyond contract execution and address broader long-term owner benefit. When owners choose to not participate fully in partnering and frown on any efforts outside the prescribed scope and contract, then it is true that the partnering group may need to stick close to simple contract execution. When working up a series of breakthroughs outside of the current project specification is clearly not going to be received well, then efforts by the EPC team to improve the owner's situation are wasted efforts and are not in line with extraordinary accomplishment. It should be clear that when owners fully participate in partnering, the potential for improvements is then a multiple of what it is if the EPC team has to stick only to the contract at hand.

Increasing the strength and life expectancy of a critical process in a facility could lead to a significant improvement in long-term economics, but is the owner willing to increase the budget, add a little to the EPC fee, and accept a three-day delay in exchange for the long-term economic gain? Understanding and applying the scoreboard measures involves a considerable amount of owner involvement. It also involves figuring out a lot of things as the project moves forward. Partnering does not expect or need to have all the scoreboard measures figured out at the beginning. It starts with what is available and develops what is needed when it is needed. Even when elaborate and reasonably reliable performance measures are available at the onset of EPC work, they often need to be revisited and revised to be able to handle breakthrough thinking.

Breakthroughs Rest on the Shoulders of Shortfalls   Partnering involves doing the right work well, and it involves taking time for creative thinking and chasing after possible breakthroughs. When the right work is done well and the breakthroughs occur, celebration comes easy and often. However, the celebrated glory comes on the backs of those who struggled through great ideas that came up short and attempted breakthroughs that just didn't quite make it. Mature partnering groups recognize all efforts that contribute to extraordinary accomplishment. These groups recognize and celebrate the shortfalls as being the stepping stones for the breakthroughs. The successes get the most press and the biggest rewards, but the extraordinary shortfall efforts are not ignored or forgotten. They are recognized for the contributions they make in their own way to accomplishing the extraordinary.

Challenging Setting   Implementing partnering for normal EPC work in a normal setting is not a difficult choice or task. However, when EPC work is not normal and the setting is especially challenging, most executives prefer to fall back on a tried-and-true owner and EPC contractor relationship. In abnormal and challenging settings the risks are usually very high and very imminent; adding the risk of a newfangled management approach is viewed as inappropriate. On the one hand, this is sound reasoning. That the situation is not sufficiently under control to be considered normal is an indication that something in the strategy and implementation is probably out of whack. To expect out-of-whack management to be successful with partnering is probably asking too much. On the other hand, what better tool than partnering to sort through what is and is not happening well and chart a new course to extraordinary accomplishment.

Implementing partnering in an abnormal and challenging setting can be extremely effective, but also extremely difficult. It is absolutely no place for beginners or so-so partnering performers. Owners faced with such a situation would want to bring an experienced partnering group on board including an experienced facilitator to work with the owner and partnering group. Because of the interpersonal base in partnering, more valuable than experience would be an existing team that can move in essentially intact. Key questions for the owners would include: How much is at risk, how much can be saved, and what would you be willing to pay to set it right?

If an owner is facing extensive and evolving legal, environmental, and market issues, choosing partnering would be a wise move. With its head-up view and steadfast focus on extraordinary accomplishment, the partnering approach is generally more effective in pulling in the information and getting the right people on the right issues as fast as possible. In fast-moving complicated situations it is often established policies, practices, and protocol that hinder getting the right people together fast enough and in the right way to get ahead. Partnering's willingness and skill in dealing with such situations are extremely effective in a normal EPC setting and ideally suited for fast-moving situations.

Hiding and Defending Mistakes   Hiding and defending mistakes are basic human instincts and standard practice in most business settings. Every minute spent on hiding and defending mistakes is a minute that could be put into assessing and resolving the situation. Partnering groups are well trained in sniffing out these situations and coolly working forward to what is in the best interest of the best outcome. Not only is the cost of the mistake minimized, but the resolution of it costs less and is more speedy. It is not that partnering has any exclusive ownership in minimizing the hiding and defending and pushing for the best resolution. What is important here is that partnering simply accepts the hiding and defending as something that happens. For extraordinary accomplishment to happen, the hiding and defending need to be discovered and dealt with in the best interests of the project.

Partnering's single-minded focus on extraordinary accomplishment causes mistakes to be viewed more as a part of reality that must be resolved than a stigma to be hidden. Outside of partnering, those people who hide and defend mistakes tend to fare better than those who are more open. Under partnering, fessing up to mistakes is still discomforting, but the rewards of opening up to mistakes far outweigh the consequences of hiding and defending them.

Partnering Tools   Partnering is blazing a few new trails in the interpersonal and inter-company cooperation area, and it is blazing a few new trails in how it looks at and deals with projects. But for the most part the tools of partnering are not new. What is new is how these tools are viewed and used. Such things as constructive confrontation, conflict resolution, creative thinking, team building, reengineering, business strategy, and the like are simply packaged and applied in a different way. The expectation to use whatever tool is best suited for the situation is more obvious, and the no-nonsense approach to dealing with anything that is in the way of extraordinary accomplishment is more obvious.

In one sense there is nothing especially extraordinary about partnering. It is part TQM. It is part MBO (management by objective). It is part MBWA (management by walking about). It is part JIT (just in time). It is part team building. It is part creative thinking. It is part problem solving. It is part decision making. It is part reengineering. It is part zero defects. It is part P-D-C-A (plan-do-check-adapt). The wonder of partnering isn't in its parts. The wonder of partnering is that it brings all these good practices together in an open and trusting work setting in such a way that they work well together.

As management fads, these tools were the centers of attention in their time. They were the thing that had to work so that other good things would happen. They all work to some extent, and they often help us accomplish major breakthroughs and results. The genius of partnering is that for once these practices are not the end, they are not even the means. And they are not the answer. The focus is not on them; the only focus is on how they can best play into extraordinary accomplishment.

If it works and it helps, it gets used. If it can be made better for the situation at hand, it is made better. If it is selecting one practice and not another, fair enough. If it is a blend of several approaches, fair enough. Because the only focus and the only criterion is extraordinary accomplishment, everything is fair game. If an approach isn't the best choice in a given situation, no one is going to force it. At the end of the day, the genius of partnering is in its single focus on the mission and its single criterion of extraordinary accomplishment. Practices and changes to practices are fair game. Regulations and changes to regulations are fair game. Personal behaviors and changes to personal behaviors are fair game. Everything is fair game, so everything can work if it supports the pursuit of extraordinary accomplishment. And no management tool has to be used unless it is the best choice for the situation at hand.

Heretofore Distasteful Tasks   An interesting twist in partnering is taking heretofore distasteful tasks and behaviors, such as enforcement, compliance, checking, investigating, debating, and challenging, and converting them into honorable pursuits that in a no-nonsense way support our quest for extraordinary accomplishment. Through intense upfront training the partnering team tends to get comfortable with these otherwise uncomfortable affronts rather quickly. A key challenge is building interpersonal skills into the partnering members so they can do more good than harm with their newfound assertiveness. Extraordinary accomplishment requires that the team be assertive, but ineffective assertion is clearly counterproductive. Soon after mastering internal assertiveness, the group masters external assertiveness, a much kinder, gentler, slower approach than what works within partnering's inner circle.

No Holding Back   Good people have lots of ideas on how things and people could work better. They have lots of ideas on how we could get right to the heart of a situation or a solution, but in day-to-day business they hold back because they might offend someone or make them uneasy. They hold back because they might make someone look bad. They hold back because if they try to get involved they feel like they will be out of line and criticized for not paying enough attention to their own work. They hold back because they don't think it's right to nose into a situation unless they have an invitation to do so. They hold back because if they spend time helping someone else they may fall behind in their own work. People hold back their ideas and help because they feel they will be better off holding back than jumping in.

Partnering makes an honest and earnest effort to reverse all that. If someone could help the pursuit of extraordinary accomplishment in any way and chooses not to, from a partnering perspective they have made a wrong choice. Holding back is counterproductive, detrimental, and discouraged. Partnering puts energy into seeking out good people's ideas and sorting out those that can help the most. Undertones of blame or failure simply don't fit into the picture; what counts is moving forward in the best way possible.

The self-actualizing aspects of partnering face a tremendous challenge in reprogramming people's thinking and redirecting people's behavior. But it works and it pays off. Take away all the inhibitions that cause people to hold back, and the released energy and creativity jump by leaps and bounds. The challenge in clearing these hindrances is not in spotting the problems. The cure is in working with those involved through the problems to the root causes and eliminating them from the working environment. Hands-on workshops dealing with real people and real issues are the best approach. Sending people to seminars and expecting their seminar lessons to be transferred back into the work setting is unrealistic. It can certainly be a part of a good program, but partnering is a day-to-day experience. For partnering, learning isn't something done through an off-site seminar. Learning is almost always a group problem identification and resolution session right on-site. Identifying and learning how to resolve detriments to extraordinary accomplishment are day-to-day responsibilities of the group and part of its daily work.

Transforming Self-Centeredness   One of the most powerful aspects of partnering comes from transforming self-centeredness into group-centeredness. We find in the most successful partnering groups that each participant is more inclined and dedicated to helping others than helping themselves. The result is astonishing. With each member of the group doing all they can to help others accomplish the most that is possible, each individual in turn finds themselves the focus of help and encouragement from all the other members. The free-flowing encouragement and admiration are like a drug, and each participant is getting a boost from all the other participants. As each burst of encouragement and help moves back and forth around the group, the combined effect multiplies. Everyone is unselfishly looking out for ways to help each other accomplish the extraordinary, and ironically the result is that each person seems to receive more power than they give out. Certainly each participant is putting a lot of energy and creativity into doing the best they can, but there is a magical compounding of power from group-centeredness. Whereas self-centered behavior tends to detract from the sum of the parts, group-centered behavior tends to amplify the sum of the parts.

Group-centeredness shows itself in many ways. The most obvious is in interpersonal communication between partnering participants. You frequently hear such questions as, "Is there anything else I could do to help you out with that?," and "What if you were to try and. . . ?" These are demonstrations of taking the time and showing an interest in helping each other out, whether or not it is standard or expected procedure. Even agendas for group meetings reflect group-centeredness by regularly including time for soliciting ideas or asking for help. You also find members thinking a lot about each other's tasks and calling each other at odd times with suggestions, ideas, or offers of help. The energy from person to person magnifies over and over and pushes the group closer and closer to extraordinary accomplishment.

Another aspect of group-centeredness is the flip side of defensiveness. To best enhance the performance of all the other members, each member is willing to face any detriment they may personally bring to the group. The energy and focus of the group is go, go, go. But from time to time individuals question their contributions and whether they may be detrimental to another individual or the group. In a good partnering culture you see very little defensiveness when suggestions are made. What you see is a refreshing focus on what it might take to resolve the situation and whether or not it's worth resolving. It isn't a matter of who is right or wrong, or who should accommodate whom. It is simply a matter of eliminating absolutely anything that gets in the way of extraordinary accomplishment, even if that means facing your own personal issues to benefit the group. Partnering involves a sincere interest on the part of each member to identify and correct whatever may be interfering with the group's maximum performance.

Joining Ongoing Partnering Group   New players joining established partnering groups are very much like new players joining any winning team. The newcomers have a lot to learn, they have to be committed, and they have to earn their way into the inner circle. Fortunately, established partnering groups are eager to bring on board anyone who can contribute to extraordinary accomplishment, just as is the case with the really great winning teams. To capitalize on the newcomers as much as possible, the established group moves quickly to earn the respect and trust of the newcomers and get them up to speed as rapidly as possible.

It is truly ironic that joining a so-so team is almost always more difficult than joining a winning team. The so-so team seems to require newcomers to "pay their dues," "show some respect," and "become a team player" (loosely translated to mean make the old-timers look good first). Newcomers to partnering don't have to waste time and energy on such nonsense. They are expected to line up behind extraordinary accomplishment as quickly and effectively as possible, and they are given as much encouragement and support as possible to improve their chances of succeeding.

Personal Biases Are Out   Partnering's contribution to the cause is held above all else. Senior member, newcomer, associate, or anyone else, it doesn't matter who or what you are; what matters is whether or not you are doing your best to accomplish the most for the common good in pursuing the mission. Personal biases, prejudices, likes, and dislikes are all dwarfed by the drive to accomplish the extraordinary. We see it over and over. When those who come into close contact with partnering see the light, they are hooked. The spirit and the rush to great accomplishment is addictive and all consuming. Fortunately, becoming fanatical about extraordinary accomplishment usually turns out to be a good thing!

Trust Is Partnering's Lifeblood   Along with all the good news about partnering comes some bad news. Partnering is a charged-up culture that is absolutely focused on noble accomplishment. It strives to accomplish the most with the least as quickly as possible and accepts whatever reality throws at it as just part of what partnering is about. You would think it couldn't be stopped, but it has an Achilles heel.

Trust is the lifeblood of partnering. If you want to destroy partnering, strike at the trust. Part of the partnering process is to stay on top of what is right and wrong with the group and prevail. Injure trust a couple of times, and partnering will come back swinging. But damage trust a little too often, and partnering will lose its balance and begin to fall. Partnering participants can accept a few trust setbacks. However, if personal communications are abused, subversive behaviors prevail, or the group is unreasonably forced to fall in line with standard practices and relations, trust in partnering can be broken and the spirit of partnering will fade. If there is a total meltdown of trust, rebuilding it is every bit as challenging as building it in the first place. Those of us involved in partnering must always be alert for trust infractions and take immediate action to resolve or cope with the situation.

Continuous Learning   Progressive learning is a way of life within partnering. The minute a practice or process or procedure is completely understood and matter-of-fact, it is ordinary. Anyone can do it, and that's not what partnering is about. Partnering's passionate pursuit of extraordinary accomplishment means that without exception the participants are always looking for a better way. Sometimes the better way is just a *little* better or faster, but it is still better. And sometimes it is a breakthrough, a cause for celebration. But the pursuit of the next lesson is at the core of partnering's success.

Partnering involves continuous improvement in strategy, approach, methods, practices, and procedures. Partnering does look for continuous improvement in everything, but it is on the lookout for more than continuous improvement. Continuous improvement suggests that something is there and that something can be improved. Partnering takes continuous improvement as what you do when you can't accomplish a major breakthrough using something entirely new. Continuous innovation is partnering's first choice, continuous improvement is its second.

The Key Players   For partnering to accomplish the most, all key players should be involved from the start. Ideally, this would include representation from the owner, EPC contractors, key subcontractors, future operations crews, future maintenance crews, regulators, future suppliers, and future customers. With all the key players represented up front, the most good can be done for the owner. Extraordinary accomplishments and breakthroughs can come from creatively looking at the project from the present situation through to the eventual shutdown. For example, by involving suppliers early on, new ideas can be put on the table to reduce costs or improve reliability. Also, by looking far into the future at the facility shutdown, major costs for decommissioning or environmental cleanup might be significantly reduced with little or no upfront costs.

When Key Players are Not Available   However, many key players may not be known at the beginning of the partnering effort. The owner is then faced with several choices. First, push ahead and let the existing players do what they can to represent the future players until they are selected and can be brought into the group. Second, contract for the services of those who are involved in such services now and have them represent the views of the future players. Third, select as many key future players as possible at the start of the project and engage them in the partnering effort well ahead of when their work will actually start. Fourth, based on letters of intent, line up and involve key players now but proceed with the understanding that formal competitive bids will be required in the future. In this case the players with the letters of intent are gambling that they will be able to win the work later, and the owner is gambling that it will not run afoul of unfair practice claims by those who lose future bids.

When players are brought on board for partnering well ahead of their true start dates, there are questions about who should absorb their costs. Should the owner pay because it is primarily for its benefit? Should the future contractors absorb the costs as part of their future work and billings? Certainly it can go either way. When future players are represented by stand-ins who are not assured of gaining any downstream work or benefit, clearly the owner (or perhaps the EPC contractor) will be required to pay for their services. The benefits of involving all the key players are usually significant and dwarf the costs, but absorbing the costs still has to be equitably worked out.

The costs of involving key players early in the partnering process can be offset by avoiding the costs and interference of a long series of bidding cycles. By selecting a full set of key contractors one time early in the process, owners and contractors alike can avoid considerable downstream costs and get the maximum value from their partnering effort. In full turnkey situations the burden is already shifted to the EPC contractor. When major equipment suppliers cannot be selected until later in engineering, then bringing key engineers in from the most likely equipment suppliers may be the best way to accommodate partnering's early brainstorming and planning sessions.

Risks and Rewards   Sharing of risks and rewards is intuitive and seems rather straightforward. Sensible people sit down, review the facts, and work out an equitable reallocation. However, sharing of benefits can get very sticky, especially when great ideas come from individuals of one entity and create significant benefit for another entity. Capitalizing on the benefit requires coming up with the idea, developing the idea, working out the details, implementing the changes, taking the risks, staying the course through the tough times and setbacks, and maintaining the improvements to realize the greatest possible benefit. In most cases each component of capitalizing on the benefit involves different people with different views on just how important they each feel they were to the final benefit. There usually are no single views of how to allocate the risks, costs, and benefits across all the components, and there are even great differences of opinion regarding the sizes of the marginal costs and benefits. Each partnering group obviously must work out a risk/reward process that suits them and one they can trust to come to a reasonable answer. In a fair allocation, each person involved in each component of capitalizing on the benefit will probably feel that they weren't quite recognized enough. Each will feel that they came out a little short. If that is where the allocation comes out, it is probably about as equitable as it can get. Allocating risks, costs, values, benefits, and rewards does not generally lend itself to logical measures, so differences of opinion will exit. Fortunately, within partnering hidden agendas and grievances quickly fade because concerns and issues are easily raised and resolved. Even when differences are fairly significant, the focus on extraordinary accomplishment tends to dwarf such differences and pulls the group back into full alignment.

There may come a time when EPC contracts will start with something like, "Whereas the owner wishes to maximize the life-cycle benefits of the proposed work to the best advantage of the owner, and whereas the EPC contractor wishes to add the most possible life-cycle value to the owner, now therefore we set down this contract outlining a sharing of risks and rewards to accomplish that end." However, defining the terms and conditions for such a noble and long-term perspective have yet to be developed.

"Contractualizing" We have bits and pieces of partnering and alliance agreements that are doing an increasingly better job of capturing what we are talking about, but working out the phrases, definitions, and algorithms that can qualify as a legal contract for the most part eludes us. In the meantime, and perhaps for all time, contracts will be contracts, and side agreements will be side agreements. Perhaps it is best to leave much of what partnering and alliances are about in the informal agreement arena.

Partnering and alliances are concepts involving good people with good intentions working things out as they move forward toward mutually beneficial goals. We may find that over-formalizing and over-legalizing such things simply will not work well and will not be in our best interests. In the end, time spent "contractualizing" partnering and alliances may be better spent getting on with what it is all about. With each deeper level of contracting we take away the flexibility and freedom to simply do what is best to meet the goals. Each additional clause brings clarity and definition, and also restriction and compliance.

As it stands today, the most efficient path seems to be to put the intent, boundaries, and oversight agreements into contracts, and leave as much as possible up to informal agreements and procedures that can evolve and be optimized as the work moves forward. If the goals are clear and all eyes are set on extraordinary performance, then the result will be as good as it can be. With good partnering and alliances, contracts themselves are called into question when they begin to interfere with optimum performance. Thus the partnering and alliance agreements may have much more to do with the final outcome than the stipulations in the contract. Along with a transformation in how we approach and complete projects we are finding a similar transformation in the intent and content of contracts.

Budgets and Schedules One of the most mind-bending aspects of partnering has to do with goals and targets for the EPC work. At the core of partnering is the relentless pursuit of extraordinary accomplishment, doing the absolute best possible regardless of circumstances. In its purest form, an owner partnering with an EPC contractor would simply request that the EPC firm do whatever it sees fit to best increase the long-term value of the owner. There would be no budget and no schedule. There would not even be a contract for a specific facility. Acting on the owner's behalf and in the owner's best interests, the EPC contractor might or might not build or expand a facility. It might rush to quickly put a small facility on-line followed by a larger facility, or take the time needed to put only one large facility on-line. In the purest form of partnering the EPC firm would work with the owner's capacity, marketing, and financial staffs to develop and pursue the best options.

This example of partnering in its purest form goes well beyond the trust levels we have in the marketplace today. However, it does help to introduce our next point. The schedules and budgets set up at the beginning of a project are best-guess numbers given what is known and what is expected regarding the project at its inception. If perfect information were known, then the schedule and budget would reflect the best possible outcome—i.e., they would represent extraordinary accomplishment. Any shortfall in the schedule or budget based on perfect information would be a step back from ideal performance. However, the schedules and budgets at the beginning of a project are only representations by those involved at the beginning of the project. Given all that the project will in reality face, perhaps the schedule and budget would be impossible to achieve. Given the lucky breaks that might happen for the project, perhaps the schedule and budget could be easily beaten. Imbedded in the schedule and budget may also be inadvertent or deliberate misrepresentations.

For decades the contracted schedule and budget have been the focus of our attention. To be realistic, partnering must acknowledge the planned schedule and budget. But to partnering they are only stakes in the sand. Partnering's only goal is extraordinary accomplishment. Spending too much time analyzing, debating, and defending budgets and schedules can detract from doing the absolute best job possible. Tracking and reporting are a part of reality. However, partnering demands an engagement by all parties to stay focused on exactly what needs to be done to accomplish the best outcome. Executives who drag the team into traditional, laborious hashing and rehashing sessions of why the project isn't working out according to the original schedule and budget can seriously derail the enthusiasm and energy needed to stay at the forward edge of extraordinary accomplishment.

At the end of the day the partnering group expects to have beaten the budget and schedule; that is their goal. But if at the end of the day the absolute best effort has not met the original schedule or budget, then what? In most circles we consider the project a failure, unless there are clear reasons why the original budget and schedule were flawed. We agree that owners and EPC contractors must use the intelligence gained from schedule and budget tracking to steer the project toward extraordinary accomplishment. But in the spirit of partnering, executives who traditionally have forced the best people to spend hour upon hour in budget meetings may need to reallocate a significant amount of that time to brainstorming sessions about what can be done to improve schedule or budget performance. Trusting the partnering team to do the best possible under the circumstances that develop is asking a lot of those executives who have heretofore ruled by the budget. However, spending less time wrestling with budgets and more time identifying and solving problems and chasing opportunities is important to partnering.

Returning to a sports analogy, champion teams intently track their statistics and constantly challenge themselves to do better. But the statistics are a tool and a measure. Spending some time on statistics is good; spending too much time on statistics detracts from time spent on practice and getting the job done. At the end of the day, it is playing your absolute best under whatever conditions you face that is the real measure of champions. The statistics are important, but extraordinary accomplishment is the only real focus. The same balance between statistics and doing is true with partnering. Too little discipline to the budget and statistics is bad, but too much is also bad. Part of the challenge of partnering is figuring out the right balance for the job at hand.

Launching Partnering   In some ways partnering for the first time requires a leap of faith. It is a little like thinking about getting in shape. You know you will have a better overall life if you do it. You can list the advantages of being in shape alongside the disadvantages of not being in shape, and there is only one logical answer. You see people around you who have gotten in shape and are obviously having a better life than you. You know it will be a little tough, at least at first, but you know you can do it if you set your mind to it. Go back through those sentences and in place of thinking about getting in shape, think about partnering. The issues are very much the same. If we can unlock some of the secrets to getting our bodies in shape, perhaps we can unlock some of the secrets to getting people committed to partnering.

So what is the key, why do some people take that leap of faith and get on a campaign to get in shape? Why is it that so few reap the rewards and so many do not? Why is it that most just stumble along their way spending valuable time and energy rationalizing why they will or might start their program tomorrow or next week, but not today? The answers to these questions also hold true for getting into partnering.

True Belief   It is not the answer to one question that triggers a commitment to launch and hold the course. Getting launched requires a mixture of answers and steps, and staying the course requires more answers and steps. A number of things have to fall into place and stay in place in order to successfully get in shape or get partnering going. Foremost in each case is a true belief that you will be better off for having *pursued* the program than not. A close second is a belief that if you succeed in your campaign you will be better off for having *succeeded*. Please notice that there are two important points here. The first point centers around the implications of starting and pursuing your program, and the second centers around what possible difference it might make in your life if you stay the course and succeed.

Believing that you will be better off for having pursued a program than not is a very individual matter. To some the only important issue is the eventual outcome, so the pursuit is not the issue. For others the impact of just pursuing the program can be seen as so negative as to outweigh any possible long-term benefit. For these people the pursuit is too costly. For getting in shape, they think of how their day-to-day life might have to change, how new schedules and behaviors might affect their friendships and family relations, and what people in general might think of them as they work their way through the program. Notice that these personal costs are not monetary; they are costly because of their impact on existing or perceived relationships. These same personal costs apply to partnering.

Jeopardy to Existing Relationships   If the person believes that their existing relationships might suffer because of the attention paid to the new program, their commitment to move forward is seriously dampened. If relationships are the most valuable aspect of their life and they fear damaging them, then they are trapped in sustaining the status quo. No matter how inviting the benefits of future relationships as a result of the program might be, they just can't endanger their present relationships to move forward. We are referring here to interpersonal relationships because they are the most common forces against behavior changes. If pursuing the new program might involve significant embarrassment, failure, or disruption to existing actual or perceived relationships, than many people will choose to not get involved, or at least not seriously involved. (In reality, when these people find themselves in a new program in spite of their resistance, they almost always find that their concerns and fears were not well founded. But that only comes with hindsight. Until proven otherwise, the concerns and fears are real and are a major inhibiting factor.)

Getting in shape involves most all of a person's relationships. Inhibiting relationships in partnering are mostly work-related. If a person is known as a hard-nosed know-it-all who gets the impossible done in spite of other people and the system, they may enjoy that image and reputation. At some point these people enjoy their leader-of-the-pack reputation more than what they are known for accomplishing, and they seriously fear losing that stature by becoming a team player and "disappearing" into a charged-up partnering group. Whether or not they think they could get more done through partnering is not the issue. Their survival and worth are in their image, and surviving means keeping their image alive and well. They will choose to protect their image and reputation over partnering because their image is what is most important to them.

"Marginals"   Some people are very insecure and will not pursue a get-in-shape program regardless of the long-term benefits because they feel such a major change could jeopardize what they have in the near term. Some business people find themselves in the same dilemma regarding partnering. Whether it is true or not, they believe that they are marginal at best in the eyes of those they work with. Jumping into a new program would mean learning new ways (meaning they have to start from ignorance) and trying new things (which they at first would not do well). To pursue a new program might mean saying stupid things, embarrassing themselves sometimes, and worrying that others might do better at partnering, which would put those that believe they are marginal in even greater career jeopardy. Given their belief that they are marginal at best in the eyes of their subordinates, peers, associates, and bosses, they are desperate to not rock the boat and take a chance of being laughed at or ridiculed. Obviously not jumping on board could do them more harm than good. But that is a hindsight issue. As they look at partnering, their concerns and fears in the near term are the most real and they hold back accordingly.

The person looking at getting in shape may be worrying about what people will think or say about them and what that will do to their relationships and image. People looking at the straightforward, open communication and sharing aspects of partnering may be worrying about what people will think or say about them and what it will do to their relationships, image, and career. If they are insecure in their career, partnering may simply look too risky. They may choose to say the right words, even join in if they have to, but their loyalty will be to personally surviving and avoiding anything that might make them look bad. As with those who eventually join a fitness program, these people who later find themselves in partnering anyway usually find that they are better off for the effort. This is not an ideal world and there will be those who are image-bound or fear-bound and cannot break free under their own power. However, with any luck some of these people will be forced to move ahead into partnering and may later appreciate the shove.

Pain and Pleasure   Putting image and fear aside as inhibitors to getting started, the next most important consideration has to do with life-cycle pain and pleasure. If those looking at getting in shape weigh the discomforts and costs of getting there against their faith in what they really believe will and will not happen downstream, the faith in future benefits is often not enough to tip the scales to action. We tend to give disproportionately more weight to pain, discomfort, and risk than we give to pleasure and benefits. With few well-noted exceptions, somehow possible pain, discomfort, and risk seem to be a little more likely and a little more real than the possibility of pleasure and benefits. That is a characteristic that served us well in our precivilized evolution, and overall it still serves us well. It is just part of our genetic instinct, and in the real world it carries more weight than coldhearted logic. The bottom line is that long-term pleasure and benefits need to significantly outweigh short-term pain and risk or we will have a very hard time generating sincere interest in moving ahead.

People on the doorstep of a fitness program are often inundated with talk about how the program has helped others and how others are now enjoying better lives. But the person standing at the door looking in probably has serious doubts as to whether even half those dreams will come true for them. So they can believe that it is a good program for a lot of people, but they may not yet believe that it will personally be a good program for them. In a similar fashion, when we consider partnering we are almost always hammering away at how much better off the project will be. However, the project does not do partnering; the participants do. Along with emphasizing the benefits to the project, we also need to emphasize how those who participate in the partnering process will personally be better off. For example, the new skills can serve the participants well in all areas of their business and personal life. Shedding old habits can make future work more productive and interesting. If there are no special monetary rewards to be earned from the partnering effort, the new skills should still lead to higher lifetime earnings. And if participation can lead to significant monetary gain, so much the better. The point is, individuals at the doorstep of partnering must feel that the long-term benefits to them personally will outweigh the short-term struggles that they will face. They have to develop a faith that after stepping into the unknown and learning new materials and concepts, after stumbling the first few times with new behaviors and taking a few missteps, that it will get better and for them personally it will end up being worth the struggle.

A Little Admiration Helps   What more can there be to jumping into getting fit or jumping into partnering? Our first considerations had to do with thoughts and beliefs. These next considerations have more to do with action. Those who are successful in a major behavioral change often make it through the process because other people knew what they were doing and knew how well they were doing. In most cases a person getting in shape will stick to it because their children are rooting for them, their associates at work are supportive, or a camaraderie is quickly established with others in the program. In other cases individuals have simply recruited someone to track their progress, a trainer, friend, or parent. There seems to be something in us that drives us to excel to get others' approval, so we can use that to our advantage.

Not all support groups are so formal. A little admiration, even from strangers, can be an impressive motivator. Others start on a program and someone they dislike laughs at them and says they will never make it. The drive to make the scoffer eat their words can push a person to succeed. In partnering these support group issues still hold. They are built into the process. Partnering energizes itself and continually pumps itself up. Most participants do receive support from others outside the partnering group, but the group itself fills the basic need for measuring up to someone's expectations. But still, being recognized for what you are doing and gaining a little outside admiration here and there (especially from a mentor) can do wonders for progress.

One Step at a Time   Whether it is getting in shape or partnering, it is done one step at a time. For getting in shape, exercising is a step-by-step issue. The thought crosses the newcomer's mind to go for a walk, and to be successful they must immediately stand up and put on their walking shoes. The trick is not in going for a walk. The trick is in building a reflex to immediately move on the thought. Pausing for more than a second will almost surely mean that you will come up with a reason to not go for the walk, at least not just then. That little voice in your head will say, "It will be better to wait and go in an hour or so," and it won't happen. A few one-second delays and days will go by without taking a walk. A pattern of failure will build and the program will be lost. The exercise battle is won or lost in the one second between thought and action.

The one-second thought-to-action battles exist in partnering, but they are easier to win. New participants in partnering know they are supposed to air concerns when they have them and get them out on the table. But in the beginning when a concern comes up for a new participant, a little voice usually says, "There will probably be a better time a little later; wait to bring that up." And after passing over several opportunities to bring up the concern, it simply doesn't happen or it finally happens but valuable time has been lost. Right from the start the partnering group works closely with each participant to help them step right in with concerns and ideas. Facing the uncomfortable is not so difficult the third and fourth time around, so getting from thought to action through the first few times is key. Like going for a walk for better health, once the new behavior has been practiced a few times it gets easier. At some point the new behaviors become second nature and the battle is won.

Right Program, Right Way, Right Time   One last point about getting in shape and getting into partnering. There seems to be no one best program of diet and exercise for everyone. Some people require a very rigid diet, and others can drift along with general vague guidelines and do just fine. Some require absolute schedules and routines for their exercise; some just wing it week to week. Some can fall into a set exercise regimen and stick to it for years; others need a variety of challenges that change almost weekly. Generally speaking, any approved diet program will work, and any sensible exercise program will work. The secret is not in the program, it is in connecting the right program to the right individual in the right way at the right time. The two most important factors are how suited the program is to the person's lifestyle and whether the program and the desire come together at just the right time. To successfully launch and complete a program to get fit requires a mix of thoughts, desires, actions, and opportunities that just happen to jell at the right moment. It can be accomplished by design or happenstance, but it requires a spark at just the right moment to set it off.

In many ways partnering is again like getting into shape. There is no one way to get started, no one schedule to follow, and no one pattern of behaviors that suits everyone. Partnering involves a mix of needs, styles, and opportunities that requires the right moment to jell. Words get spoken, meetings get held, exercises get done, and lists get made up. But partnering requires its own spark of sincerity and straightforwardness. And from that spark the momentum builds. Going through the motions of partnering at some point starts transforming into proper partnering behaviors and results, and eventually the participants internalize the whole process and reach new, previously untapped levels of cooperation, creativity, and progress. Instead of thinking about how to partner and what needs to be done later, the group's focus shifts to finding and pursuing the best path to extraordinary accomplishment. The person who gets fit and starts enjoying the better life that comes with it doesn't really think about the fitness program. They just live it. Once partnering is up and running and the participants have internalized the new behaviors and drives, they don't really think about partnering as partnering. It is just the way they are and how they work. It is no longer a partnering program; it is simply the way extraordinary work gets done.

15

An Industry Example— Bechtel Procurement—The Strategic Supplier Program

This is an example of how partnering can be undertaken in one area of a business and have a direct effect on another area of the business. In procurement we formed vendor alliances through a Strategic Supplier Program.

The procurement function is an integral part of the EPC business. It is an important area because 30 percent to 60 percent of project costs are in purchased materials, equipment, and services. What happened with Bechtel's Strategic Supplier Program is an excellent demonstration of how partnering can work down the supply chain to vendors and ultimately benefit the client. Figure 15.1 is the strategic supplier program goal vision mission statement.

In the early 1990s, Bechtel began sponsoring several internal programs of continuous improvement, waste elimination, and reengineering. The emphasis was to deliver superior performance to clients through lowest total installed cost (TIC) and shortest schedule. These programs were companywide and were intended to make Bechtel the engineer/contractor of choice.

Procurement offered a significant area under the EPC framework to reduce TIC and schedule. In 1993, procurement management began to look at ways to deliver savings to the engineering construction process.

Procurement recognized the importance of suppliers' contributions as critical to customer satisfaction. It set an objective to assist customers in obtaining the highest-quality equipment, materials, and services, on time and at the lowest total cost through the systematic enhancement of the acquisition process. Bechtel procurement proposed the Strategic Supplier Program to achieve the aforementioned ob-

---

STRATEGIC SUPPLIER PROGRAM
GOAL VISION MISSION STATEMENT

Mission

- Bechtel procurement is committed to the continual improvement of its services and products through systematic enhancement of the acquisition process. We are committed to being a leader in a quality of services we provide and to achieving these objectives with full participation of our employees, customers, and suppliers.

Continuous Improvement

- Vital to the success of our mission is a supplier base that subscribes to a compatible philosophy and is effective in implementing the principles of continuous improvement.

Supplier Relationships

- We seek to establish mutually rewarding business relationships that will foster openness and trust; promote better understanding of present and future requirements, processes, and goals; reduce the total cost of doing business; and enhance the quality of our products and services.
- We select our sources of supply on the basis of qualifications, performance, and individual project requirements. In addition, we will seek suppliers who demonstrate that they are committed to continually improving the quality and value of their products, who are environmentally responsible, and who provide safe working conditions for their employees.
- We actively pursue business relationships with small, minority, and women-owned business enterprises.
- Our goal is to maximize the business we do with suppliers who consistently meet our expectations at a competitive total cost level.

Figure 15.1.

jectives through the establishment of long-term relationships with selected suppliers. To accomplish these goals, procurement decided to enter into multi project acquisition agreements (MPAAs) that were intended to cover requirements for repetitive goods and services that were widely used by projects. Figure 15.2 details the purpose of the multi project acquisitions agreements.

A key objective of the Strategic Supplier Program was to better support our customers by expanding the Bechtel team to include our valued suppliers. Strategic suppliers were to be part of a select group of companies with whom we anticipate long-term business relationships. This, as appropriated, led to expanded opportunities for co-engineering with suppliers and included activities such as constructability reviews and implementation of materials management processes.

One area quickly identified by procurement that would offer leverage in the

---

MULTI PROJECT ACQUISITION AGREEMENTS

We are committed to lowering total project costs, shortening cycle times, and enhancing the quality of our projects and services through long-term relationships with selected suppliers. To accomplish this goal, we will enter into Multi Project Acquisition Agreements.

Multi Project Acquisition Agreements are intended to cover Bechtel requirements for repetitive goods and services widely used on Bechtel projects in diverse geographical locations. The agreements will combine the benefits of our traditional blanket purchase orders with our continuous improvement philosophy.

Multi Project Acquisition Agreements are non-exclusive. When possible they will be used in lieu of competitive bidding to streamline the acquisition process and reduce cycle time, better utilizing Bechtel and supplier capabilities and resources to reduce material cost.

Figure 15.2.

procurement process was the quantity purchase of common "commodity" type items. These were items that are ordered by almost all projects in large quantities, but are of relatively low individual dollar value. Procurement recognized that TIC was the overall important cost to clients, but clients focus first on the price point, and price became an almost insurmountable issue if it were not the lowest.

After much internal discussion, procurement management chose to focus its efforts on commodities that represent a significant portion of the overall project cost. In construction, the largest commodities are pipe, valves, steel, cable, hangers, and electrical bulks. Procurement decided to defer working on the larger ticket items such as tanks and conveyer systems, for example, until later.

At the time, within Bechtel each project engineering group developed its own specifications for commodity items required for a particular project. If procurement could make all the different cross-functional engineering groups agree to standard specifications on these commodity type items, larger-quantity purchases of industry common items would yield savings. Buying would then be for Bechtel, not for one project. To obtain the lowest price, procurement reasoned that Bechtel had to use the most common and standard items and that particular specifications that were not necessary only served to increase costs. Procurement approached the standards documentation with the engineering group by saying that, if a particular part of a specification was not absolutely necessary, leave it out.

In addition, the improvement of work processes both internally within Bechtel and externally with the vendors were also areas for potential cost savings. If standard items could be used, the bid negotiation cycle could be reduced, thus saving money and reducing project schedules.

A group within procurement was formed and tasked with the assignment of establishing the strategic supplier program and obtaining MPAAs. The program was interdisciplinary in the sense that it required engineering to develop and agree to standards that procurement could then use for bidding to establish MPAAs. These standards would have to be developed across functional lines. It was also expected that work process improvement would result from the definition of standards. The team was funded and ran the effort like a project with milestones, target dates, and expected deliverables.

Procurement and engineering in Bechtel both reported to the same executive sponsor that facilitated the process. Procurement senior management and the project leader spent numerous hours making presentations explaining the objectives of the program from senior management down to middle management and several of the standing discipline and office committees. The message was communicated throughout Bechtel to promote awareness.

As the standards were being developed, procurement prequalified a group of potential bidders. Some of these would become strategic suppliers and would sign MPAAs. Procurement had the potential bidders meet with the engineering groups during the standards development. Their input was quickly recognized as important, and strong relationships were built among the engineering groups, procurement, and the vendors.

One of the more important requirements for being selected as a strategic supplier was that the potential vendor have a continuous improvement program in place and demonstrate that it was working through process improvement. Supplier continuous improvement teams (SCITs) would be formed and would be composed of Bechtel, supplier, and customer representatives as appropriate, which would identify, evaluate, and improve work processes through continuous improvement. The teams were an essential part of the strategic supplier partnering evaluation and eventual arrangement.

Because Bechtel procurement had not proposed a program on such a scale before, Bechtel limited its exposure by designing the program to have non-exclusive, two-year renewable contracts with the chosen strategic suppliers. This would give Bechtel a way out on an individual project basis when it could not use the MPAAs either because of the client or the project management, and it also limited the long-term commitment.

The program funding began in October 1993 and was slated for a year. The most difficult time was during the first six months when it was necessary to change work habits and overcome basic fears within Bechtel. In September of 1994, the program team had been so successful that they returned for additional program funding to expand the program and include additional vendors and engineering functions.

After one year, by October 1994, the results were evident, and there were some direct results that could be identified:

- The project team had met its original objectives and had 24 MPAAs in place with an additional 4 pending and to be signed by the end of the year.
- Substantial savings of 23 times the program investment had been identified.
- Procurement was seeing co-engineering and earlier involvement in projects with alliance suppliers.

- A trend toward standardizing of specifications and drawing and data requirements was beginning.
- Procurement was seeing a trend to have alliance suppliers participate more in the design and select process, thereby reducing our design effort and job hours.
- More attractive and competitive pricing was being offered by alliance suppliers.
- In using alliance suppliers on a sole-source basis, procurement eliminated the bid, evaluation, and recommendation cycle of the acquisition process. Terms and conditions had already been approved by using an alliance supplier, thereby eliminating a sometimes costly and lengthy negotiation process.
- In the field, procurement was seeing certain bulk material alliance suppliers do takeoffs for Bechtel.
- Alliance suppliers were running their own project warehouses, keeping inventory, and issuing material as required to Bechtel.
- Materials were arriving at the site already "bagged and tagged." This resulted in reduced field engineer, buyer, and materials personnel involved in the supply process.
- Electronic data interchange (EDI) was being adopted with many of the alliance suppliers. This was a tremendous savings in time and labor.
- Many of the alliance suppliers have computerized pricing programs, and these are being installed in Bechtel offices with the MPAA pricing and are being utilized by engineering, procurement, and estimating.
- Procurement was able to deliver costs 16 percent below levels previously seen.

METRICS USED BY PROCUREMENT

When estimating the cost savings that were achieved by utilizing an MPAA, it was assessed that there are three main elements:

- Material price savings
- Labor cost reductions
- Schedule improvement

To facilitate companywide consistent reporting of credible cost savings data, guidelines were established as detailed in Figure 15.3.

The metrics were purposefully collected on a conservative basis to establish a credible measurement system.

Looking back, some of the lessons learned in achieving the strategic supplier program were the following:

Tangible achievements were slow in coming in the beginning. Engineering personnel had their own ideas of how to do things and saw standardization as an

PROCUREMENT METRICS

Material Price Savings

The average material savings percentage was determined as the cost difference between an individual purchase and the MPAA purchase price.

Labor Cost Reductions

Service savings was measured on the basis of reduced Bechtel job hours. The following considerations have reduced labor costs:

- Supplier providing services normally performed by Bechtel.
- Reduction or elimination of a wide range of disciplines, such as specification writing, expediting, submittal review, QA/QC, and logistics.
- Standardization of processes (enhanced productivity and reduced training).
- Reduction or elimination of support service groups (office service functions, mail, AT, project control, warehousing costs, travel costs, etc.).

Schedule Improvement

Savings associated with schedule improvement were identified as those that, as a direct consequence, reduced overhead costs due to a shorter project duration. The following considerations are included in the schedule improvement element:

- Improved deliveries through specifying standard products.
- Working with suppliers supporting JIT delivery.
- Implementation of labor- and time-sensitive techniques such as EDI and CAD.

Other areas considered in measuring MPAA cost savings which were not covered by the guidelines included but were not limited to the following:

- Leveraging of multiple project requirements not covered by global/regional agreements.
- Cost reductions related to SCIT activities with suppliers who do not have MPAAs with Bechtel.
- Shift of design, engineeering or other services to suppliers when beneficial to Bechtel.
- Reduced inventory resulting in reduced project surplus costs.
- All other negotiated values (e.g., discounts achieved through the implementation of EDI, value-added programs, incentives, etc.)
- Process improvements (e.g., reduced material pricing resulting from specification changes and/or standardization of items specified, substitution of comparable material, or process changes).

Figure 15.3.

erosion of job security. It was the vendors who eventually prevailed with examples of cost savings and reduced schedule requirements.

Procurement personnel saw the MPAAs as an erosion of the buyer function. The fear of job loss had to be addressed as a reality of competitiveness and work process improvement. Work process improvement had to be sold as the future of the company.

Client procurement had local favorites and authorship of major supplier agreements. Negotiations in many cases prevailed to use the MPAAs, but in some cases client agreements were used.

The original schedule was aggressive, and in some cases achievement of milestones did not occur as planned. It was agreed that the schedule offered an internal target for achievement. Eventually, all the milestones were delivered.

Procurement recognized the strategic supplier program as a continuous process; although the original funding has been expended, additional funding was provided for ongoing continuous improvement and work process improvement activities.

Procurement believes that the next major hurdle in the continuous improvement process is to develop standard engineering stock codes.

In the engineering construction business, procurement by its nature acted as an independent third party to obtain the best value for clients. In the classic approach to the procurement process, this independence was maintained by the arm's-length bid cycle. MPAAs represented the opposite of this approach and did away with much of the individual bidding. In the current vernacular this was a significant paradigm shift. With the Strategic Supplier Program, people are recognizing that procurement can still be an independent third party and obtain arm's-length transactions.

In summary, the program to reduce TIC and schedule by quantity buying of common commodities through establishing strategic multi project acquisition agreements (MPAAs) has demonstrated that it can be extremely successful. The partnering arrangements have all of the necessary elements—vision, mission, trust, open communication, mutual benefit, and continuous improvement—that one would find between two parties, and in this case it was the vendors who became part of an alliance that would help Bechtel with its clients.

16

Lessons Learned

Bechtel has been one of the companies leading in the use of partnering since the late 1980s. Bechtel's partnering clients have been major corporations in the power, chemical, petrochemical, and petroleum business arenas. Bechtel also has major partnering experiences with the Department of Energy and the U.S. Army Corps of Engineers. In addition, Bechtel has used partnering successfully within its own organization to establish the strategic supplier program.

Although the bulk of the client partnering work has been with maintenance and revamp projects, partnering is being used on major engineering and construction contracts. The size of the work orders has ranged from only $900 to over $2 billion with an average of a half million dollars. Personnel has peaked on some partnering projects at over 900. Individual work orders simultaneously outstanding for one client have been as high as 100.

In this chapter we document examples of lessons learned from these partnering alliances. These focus on the successful lessons learned, but some that could be classified as negative are also included because they pose opportunities and challenges for future partnering.

To be successful, the basic elements of partnering are required to be present in the relationship:

- Commitment to a specific goal vision mission
- Trust
- Open communication
- The elements of risk and reward
- Knowledge sharing

- An embracing of continuous improvement
- Metrics
- People

PARTNERING LESSONS LEARNED

Mutually Beneficial Arrangement

The most important lesson that we learned was that the partnering alliances must be based on a win-win approach whereby both parties understand and work to make the partnering relationship mutually beneficial for both parties.

Hard Work

Partnering is hard work. All the projects had ups and downs and required day-to-day effort to make partnering successful. Partnering requires the commitment and effort of many people to work correctly. When people have been able to attain a high level of partnering, the benefits have been marked and the results of the synergy have been dramatic.

Savings

The partnering alliances have demonstrable cost savings and schedule reductions that translate directly to added value to the clients. We have found that when partnering is working it is not unusual to save 10 percent on total installed cost and 10 percent to 11 percent on schedule.

Continuous Improvement

One of the basic elements of partnering is continuous improvement or what is sometimes referred to as quality management. A sample quality management program is outlined on Figure 16.1.

Quality improvement teams (QITs) are made up of the members of the partnering team and are the basis for continuous improvement programs. In our partnering, the teams benefited from the members' diverse experience and knowledge. In the nonadversarial environment, problems were discussed and solutions were openly sought. We found that when larger activities were identified, it was better to make a QIT responsible for executing the activities identified for improvement. We also found that identifying opportunities for improvement was an ongoing process in which everyone participated.

One of the alliances openly tied continuous improvement to the Deming principles and had six specific focuses. This is detailed on Figure 16.2.

QUALITY MANAGEMENT PROGRAM

Purpose

The purpose of the ABC Quality Management (QM) Program is to establish and promote a process of continual quality improvement.

Organization

The organization for QM shall consist of a QM Management Committee (QMMC), QM Steering Committees (QMSCs), and Quality Improvement Teams (QITs).

Charter/responsibilities

The QMMC will manage the overall QM program. It will be responsible for ensuring that a process of continual quality improvement is established and promoted. Responsibilities include

- Interface/coordination with Bechtel management on QM issues.
- Ensuring proper QM related training is provided.
- Soliciting/reviewing/selecting opportunities for quality improvement.
- Establishing and overviewing QMSCs.
- Reviewing and implementing quality improvement recommendations, as appropriate.

The QMSCs will be responsible for

- Developing the description for quality improvement opportunities and selecting participants for QITs.
- Reviewing QITs' efforts/recommendations and interfacing with QMMC as required to provide direction to the QITs.
- Assisting the QMMC and QITs with the implementation of approved quality improvement recommendations.

The QITs will be responsible for executing most of the activities identified for improvement. The QMSC and QMMC will provide guidance, approvals, and support as required.

Figure 16.1.

Reimbursable Work

We found that a reimbursable contract was the best vehicle for partnering work. Although most of the work has been reimbursable, some individual work orders have been lump sum, unit price, and even combinations.

ALLIANCE'S SIX FOCUSES

- **Leadership.** All members of the Alliance management teams are fully trained and are firm believers in the quality process.
- **Education:** The Alliance has a well-planned and standardized training program that covers technical and non-technical topics for all its employees.
- **Infrastructure:** An effective organization, with a Senior Management Quality Team, a Steering Committee, and various sponsors, advisors, and coordinator is in place. This infrastructure coupled with the principles and application of high-performance team technology increase the involvement, effectiveness, and productivity to the Alliance project teams.
- **Plan:** A detailed linkage of work processes has been developed. Alliance projects have established various teams engaged in the improvement of specific areas. Each team has a mission consistent with the corporate mission. All teams are client-focused and depend heavily on feedback for continuous improvement.
- **Improvement:** The Alliance utilizes a systematic approach to continuous improvement based on the Plan-Do-Check-Adapt cycle for quality improvement. When appropriate, selected data are collected and statistical process control techniques are employed to analyze systems.
- **Involvement:** The most important element of the quality process is the total involvement of all stakeholders, ABC-Bechtel, vendors, management, and employees at all levels.

Figure 16.2.

Expectations

The better defined the client expectations of how the partnering is to work, the faster the partnering process can start and the higher the degree of partnering that takes place. This experience was particularly noticeable with two of the clients who had used partnering previously and were committed to it as a way of doing business.

Work Process Improvement

When the partnering teams work, they usually have to combine the work processes of both companies. Partnering seems to encourage people to find ways to do work processes that are simpler and more efficient. There are several examples where work processes that reached deeply into the organizations were simplified, resulting in significant cost savings for the client.

Safety

On all the partnering projects, we found that safety consciousness is significantly increased, resulting in a sharp reduction in lost-time accidents. On one project we have worked over five million job hours without a lost-time accident.

Global Partnering

Although partnering has not been as successful internationally as expected primarily because of cultural differences, we found that when a current partnering client makes the commitment, partnering can be successful overseas. One client has worked with us on successful partnering projects in Russia, United Kingdom, Korea, and Australia.

Positioning

Partnering is useful as a positioning tool. We found that with trust, in some cases, the client would share future capital plans with us. This lead time allowed us to position ourselves for the future work.

Outside Facilitator

Even with teams that were familiar with the partnering concepts, we found that an outside facilitator could jump-start a partnering alliance. In all the cases where we used the outside facilitator to do high-performance team building, the benefits were evident in the more focused start-up and early relationship building that took place.

Goal Mission Vision Statements

One of the key components of the partnering alliances is the goal mission vision statements. These statements serve to focus the team members' attention on an immediate and tangible set of goals and cause the culture shift, which is one of the key reasons partnering works. A good facilitator is able to take the detailed client objectives and have the members of the team convert them into a goal vision mission statement for the alliance. Three examples of good goal client objectives are displayed in Figure 16.3, Figure 16.4, and Figure 16.5. Good examples of goal mission vision statements are displayed in Figure 16.6 and Figure 16.7.

Breakdown Sessions

One of the partnering alliances recognized how fragile the partnering process was and how much work it would require. Whenever the alliance seemed stalled for some reason and partnering was not working as well as the team thought it should be, they would have a breakdown session. The purpose of the breakdown session was to get back on track by analyzing the breakdown and having everyone work to fix it.

Owner Expectations

When the owner has clear expectations on what and how it wants the alliance partner to act, we found the partnering works better. Figure 16.8 in which ABC

ALLIANCE OBJECTIVES

Lower total installed cost of capital projects, shorter time to market, and better quality plants are the Alliance's objectives. In addition, we work to provide value engineering and constructability that will result in higher plant yield, lower operating cost, longer plant life, and an environmentally responsible and safer facility. As we continue to quantify our value-added benefits, we will remain focused on these three objectives.

Lower Total Installed Cost

Lowering the total cost for projects allows ABC to optimize capital expenditures and enhance its return on investment. The Alliance works to lower total cost by

- Accurate project estimates
- Value engineering and constructability
- Eliminating work process variation
- Integrated materials management
- Effective contracts administration
- Optimizing specifications and standards
- Eliminating waste in work processes
- Effective cash flow management
- Key supplier/contractor quality partnerships
- Effective front-end loading techniques
- "Best buy" purchasing for materials and equipment
- Cost-saving commercial negotiation
- Reduced rework
- Optimizing project man hours

Shorter Time to Market

Through shorter project schedules, the Alliance provides ABC a quicker time to market for its products. This is achieved by

- Establishing accurate project schedules
- Shortening engineering schedules
- Optimizing the bid, and reducing evaluate and commit cycles for materials
- Key supplier/contractor buy-in to shorter delivery cycles through continuous improvement
- Shorter construction schedules through the effective utilization of constructability and value engineering techniques
- Integrated materials management techniques
- Eliminating waste in work processes
- Developing and implementing standard work processes and "best practices" sharing across all projects

Figure 16.3.

Alliance Objectives *(Continued)*

Better Quality Plants

Providing a better quality plant for ABC is achieved by

- Utilizing value engineering and constructability techniques that will result in a higher yield for the facility
- Working closely with ABC operations, maintenance, and engineering to design and construct a plant with lower operating costs
- Optimizing design to achieve a longer plant life
- Integrating environmental requirements in the design
- Making plant safety a high priority during construction and subsequent operation

Figure 16.3. (Continued)

is the client, clearly details expectations. This affects the partnering by clearly communicating to both organizations what the expectations are and how the partnering alliance is to work.

Organizational Behavior Roles and Responsibilities

One aspect of a good partnering alliance is defining the roles and responsibilities of each of the stakeholders in the alliance (see Figure 16.9 and Figure 16.10). These roles and responsibilities detail generally what the two organizations are responsible for and how they will act in the partnering alliance.

In one alliance, the partnering team developed guidelines on how people would personally interact with team members. These guidelines are detailed in Figure 16.11.

CUSTOMER'S PHILOSOPHY AND OBJECTIVES

Customer's Philosophy for Suppliers

- Limit the number of suppliers to those best able to meet needs
- Establish long-term business relationships based on mutual trust
- Establish synergistic team concept and win-win attitude

Customer's Objectives with Contractors

- Use contractors to do the EPC work
- Utilize own resources to improve reliability, productivity, efficiency and profitability of facilities

Figure 16.4.

PROGRAM DESCRIPTION

ABC has selected Bechtel to perform engineering, procurement, and construction management services for multiple projects at ABC's manufacturing complex, as well as other ABC facilities.

This work is to be performed at the manufacturing complex and other ABC locations. ABC is to select its involvement in each project and Bechtel is to provide all additional services as required.

ABC has developed a list of expectations for the program that are listed below.

Basically the concept is for Bechtel to team up with ABC for the execution of designated projects and studies. ABC resources will be supplemental with Bechtel personnel to perform the assignments.

There will be multiple projects handled simultaneously with individual schedules, strategies, and reporting. Projects will consist primarily of plant-modernization-type work ranging in size from $5,000 to $50 million. These projects will generally be directed by ABC. ABC may also elect to assign larger projects to Bechtel that will generally be directed by ABC's head office headquartered in New York.

Bechtel's services will include

- Studies
- Development of scope documents
- Preparation of screening and AFE estimates
- Preparation of AFE authorization packages
- Scheduling and resource planning
- Preparation of strategies for execution of engineering, procurement and construction
- Performing engineering, procurement, and construction management services for multiple projects
- Monitoring project costs, schedules, and materials
- Preparation of operating manuals
- Spare parts planning
- Training
- QA/QC
- Safety direction and programs
- Field support including
- Verification of existing systems
- Interpreting engineering documentation for construction personnel
- Other EPCM services as requested by ABC

Figure 16.5.

THE ABC BECHTEL ALLIANCE

Vision

The ABC-Bechtel Alliance contributes to the business success of each partner. We are the preferred providers of project and technical services and the assignment of choice for our people. Our Alliance is one in the view of our customers, employees, and the industry.

Mission

The ABC-Bechtel Alliance provides innovative, value-added project and technical services that meet the ABC business expectations by

- Being low cost
- Performing work ethically, safely, and in an environmentally sound manner
- Fostering an environment for personal success and growth of our employees
- Contributing to the business success of each partner
- The synergistic participation of each partner

Objectives

- Meet customers' expectations
- Challenge each other
- Reduce TIC by 25 percent from historic norms
- Reduce project cycle time
- Provide reliable, superior technical resources to support business requirements
- Provide resources and work processes that are complementary and low cost
- Perform work without any safety or environmental incidents
- Utilize best business practices
- Apply innovation and creativity
- Learn from our lessons learned
- Measure results—take appropriate actions
- Demonstrate continuous improvement
- Promote teamwork, trust, and open communications
- Foster personally rewarding work environment
- Pay for performance by sharing risks/rewards

Figure 16.6.

GOAL, MISSION, AND PEOPLE PRINCIPLES STATEMENT

Our Goal

Bechtel is committed to understanding and exceeding ABC's expectations for performing quality engineering, procurement, and construction management and related services. We are dedicated to providing ABC with safe, environmentally sound, reliable operating facilities that meet all ABC's technical requirements and schedule objectives and have the lowest overall life-cycle cost for installation, operation, and maintenance.

Our Mission

We will be ABC's preferred choice for all EPC services by

- Being a highly proactive, cooperative, and innovative team that seeks breakthroughs in project execution improvements and exceeding customer expectations;
- Being a seamless organization providing EPC services with work processes optimized across functional boundaries through the use of integrated automated systems and networks;
- Being an industry model for an owner/contractor partnership where the win/win benefits fully satisfy both parties; and
- Providing quality and exceptional value of services.

Our People Principles

- Management recognizes that people want to do a good job but that the system often prevents them.
- Management establishes an environment that fosters open communications, cooperation, and participation of all personnel in the continuous improvement process.
- Management recognizes and rewards contributions and provides career development and advancement opportunities equally and fairly.
- Management recognizes that team members are equally concerned about overall team success and individual success and proactively help each other achieve both regardless of functional discipline alignments.
- Management recognizes that team members enjoy and take pride in their work.

Figure 16.7.

ABC EXPECTATIONS

The following list of expectations was developed by ABC and provided to Bechtel to help ensure a common understanding of ABC's objectives related to the program:

- Safety is number one.
- Seek/try to understand ABC well enough to react as ABC would.
- Accept and apply ABC specified methods, practices, and specifications.
- Assume initiative—including making tough decisions in ABC's best interests.
- Challenge ABC instructions—when they appear in conflict with other ABC expectations.
- Be team-oriented—especially internally.
- Be proficient and cost-effective in all aspects of the work.
- Be open and objective about execution mistakes.
- Provocatively manage in ABC's best interest.

Figure 16.8.

PRIMARY ROLES AND RESPONSIBILITIES

Owner	Bechtel
• Negotiate owner/regulatory commitment	• Prior buy-in and commitment
• Interface with the stakeholders	• Active support and participation
• Exercise financial and managerial control	• Develop baselines, trends, and change control input
• Negotiate and approve work strategies, establish success criteria	• Develop and recommend cleanup strategies; perform the work successfully
• Assess Bechtel team performance	• Provide self-evaluation
• Jointly responsible for results	• Win together, lose together
• Ensure timely review/comment/approval of deliverables	• Submit quality deliverables on schedule
• Enable team to implement procurement system	• Tailor procurement system for work
• Eliminate impediments to success	• Identify processes, standards, and procedures that impede success
• Keep the contract current with the work	• Adhere to contract terms and conditions

Figure 16.9.

ORGANIZATIONAL ROLES AND RESPONSIBILITIES

The Bechtel Team Can Be Counted on for

- Controlling cost
- Adhering to schedule (baseline)
- Having technical excellence
- Adhering to contractual terms and conditions
- Maintaining a safe workplace
- Taking technical and managerial leadership
- Recommending direction and approach
- Challenging the status quo
- Inputting and buying-in to ABC decision process
- Developing innovative approaches (across the board)
- Following through on commitments
- Directing support in ABC's interaction with stakeholders
- Supporting ABC's transition into project orientation
- Having high ethical standards
- Being proactive in long-range planning process
- Performing objective self-evaluation
- Seeking appropriate balance between risk and reward
- Supporting ABC in telling project success stories
- Being steadfast when necessary, flexible when appropriate
- Involving and not surprising ABC in interactions with headquarters
- Keeping commitment to the project commitment statement
- Working cooperatively with other site contractors

ABC Can Be Counted on for

- Being fair/honest/objective
- Making timely decisions (agree to timely)
- Ensuring strong teaming continues
- Challenging unnecessary requirements—support Bechtel recommendations
- Being a filter for outside demands
- Communicating effectively
- Creating an atmosphere of empowerment for project managers
- Being accountable for ABC decisions
- Holding Bechtel accountable for their actions and reward performance
- Maintaining our commitment to commitment statement
- Not micromanaging
- Working hard and having fun
- Fulfilling our roles and responsbilities
- Being open, learning from Bechtel, being receptive to new ideas
- Having high expectations

Figure 16.10.

TEAM GUIDELINES TO LIVE BY

Ground Rules

We Will Demonstrate Mutual Respect

- Don't complain or discuss hearsay.
- Lead by example.

We Will Treat All Members as Shareholders

- Respect the expertise of each member.
- Seek support of the other members.
- Offer assistance to the other members.
- None of us wins unless all of us win.
- Clarify to each member the common objectives of the assignment.

We Will Share All Information with the Other Members

- Strive to understand the other member's business and operating practices.
- Use the information only for the good of the team.

Figure 16.11.

Choice of Clients

We found that it was important to pick our alliance partners carefully. One of the key criteria for having a good alliance is that both companies have active continuous improvement programs. A continuous improvement program is an indicator that a win-win relationship will develop. When the partnering relationship fails, the culture of the two parties usually reverts back to following the terms and conditions of the underlying contract, and the trust and openness diminish to an adversarial relationship. The proper choice of the alliance partner is critical for a successful alliance.

Choice of Partnering Team Members

Just as important as the choice of the right partner is the choice of the right team members for the alliance. Generally speaking, we found that the relationships that were built in the alliance required people who were not just technical experts, but team members who had both strong interpersonal skills and technical expertise. The give and take and the honesty and openness of team relationships required team members who could participate in the discussions and contribute to the problem-solving sessions. The day-to-day contact with the client personnel of all levels required team members who were generally extroverts.

Partnering with the Government

The government is beginning to strongly support the partnering concept. We have seen the U.S. Army Corps of Engineers direct that partnering will be used whenever possible on projects because of its success in avoiding disputes and claims. The Corps qualifies partnering by making it voluntary. We have also seen other agencies of the government use partnering on longer-term projects and where there is a specific reason such as a need to change an inbred culture that existed on the job site.

KEY METRICS AND MEASURES

Capital Effectiveness Measures

- Cost versus appropriations
- Engineering construction schedule slip
- Attainment at months end
- Start-up time
- Cost versus schedule
- Cost savings

Duration Reduction Measures

- Average project cycle time projects > $5 million
- Average project cycle time projects < $5 million
- Average project cycle time small capital projects
- Systemic process changes—days saved

Behavioral Characteristics Measures

- Customer rating
- Employee rating

Safety Measures

- Lost time accidents—construction
- OSHA recordable rate—construction
- Lost time accidents—engineering
- OSHA recordable rate—engineering

Figure 16.12.

Metrics

Metrics is the measurement that provides feedback to the team members on how well they are doing in meeting their specific goals. Metrics in partnering usually focuses on four areas: capital effectiveness—cost; duration reduction—schedule; behavioral characteristics—client satisfaction; and safety. Each alliance team, and then each work order team if it is for a larger work order, establishes metrics. The metrics are usually agreed to by all parties in the partnering relationship. Measurement is usually done on a monthly basis for some items such as cost, schedule, and safety. Others, such as client satisfaction, may be done monthly, but in some cases they may be done more extensively on a quarterly basis. A typical set of detailed metrics are displayed in Figure 16.12.

A tracking system is used to measure performance of individual items. Variable point scores are set for each item, and weighted totals are calculated and summed for each major category. The scores are then reviewed each month as part of assessing the alliance's overall performance.

Incentives—Sharing the Rewards and the Risks

In many of the partnering alliances there is a provision for sharing risks and rewards. This is the incentive portion of the fee, which is put at risk by the engineer contractor, and the amount awarded depends on performance as measured by the metrics. In some cases, the award can be increased by using prior savings in cost or schedule, thus giving the engineer contractor more fee for better performance.

In some of the more progressive alliances, a certain percentage of the awarded incentive fee is paid to the employees. This is a way to monetarily reward behavior on a quarterly basis. The basics of one program are outlined in Figure 16.13.

Management Involvement

We found that senior management from both the companies have to be involved from the start in all partnering alliances. The participation begins with the first high-performance team building session. Senior managers from the companies in the alliances participate in these meetings.

After the partnering alliance has started, the sponsoring team of senior management from both companies meets periodically, usually no less than semiannually. The agenda for the meeting usually includes reviewing the alliance performance, general business conditions, and opportunities for improvement.

Along with the sponsoring team of senior management, there is usually a steering committee of the alliance management and the next level up of management in the respective organizations. This steering committee usually meets monthly to review the alliance's performance, but focuses on individual work order or project performance. Sometimes a special topic is discussed that focuses on some external aspect of the alliance.

EMPLOYEE INCENTIVE PROGRAM

Purpose

The purpose of the Employee Incentive Program is to share project incentive awards with Bechtel employees involved in the work on the ABC Program in order to encourage greater teamwork and commitment to achieving project incentive goals.

Project Incentives

Project incentives will be established in accordance with contract terms and as mutually agreed between the responsible Bechtel and ABC representatives.

Employee Incentive Pool

A certain percent of the actual Bechtel Incentive Project awards will be allocated to the Employee Incentive Pool.

Employee Incentive Awards Disbursement

- The employee incentive pool will be disbursed quarterly to all Bechtel employees who worked on the ABC Program during the established qualification period.
- The requalification period is the six-month period ending the month incentives are awarded.
- The employee incentive pool will be disbursed to employees who worked a minimum of 200 reimbursable straight-time regular hours on ABC's Program during the qualification period.

Figure 16.13.

The two program managers or directors of the alliance meet almost daily. They work on the day-to-day operations of the alliance.

Use of Initiatives

We found that one way the alliance partners can participate in a win-win strategy is by making small investments (sometimes called initiatives) in non-project-related alliance activities where an improvement profits the alliance overall and not just one project. The investments are an exhibition of commitment. They add value, improve work processes, build the relationship, and are viewed as a positioning for future work. These investments, for example, can involve using automation integration, such as the integration of estimating and engineering work processes for the benefit of all the alliance members.

Down Partnering

In our partnering alliances, we found that the working concept of partnering usually stays high in the organization, but that the most successful partnering relationships were those in which down partnering was used to involve the maximum number of project people in the basic concepts and interplay of the relationships.

Flexibility Is the Key

Just as each partnering relationship is different, so are the terms and conditions of the contracts that underlie the partnering relationship. We found that each client viewed the incentive provisions and risk/reward sharing differently, and that it was important to be flexible in our approach when these provision were negotiated.

Multiplier

The multiplier is the factor that is applied to the base salaries to cover overhead on work orders or projects. The multiplier in most alliances is reviewed annually to be sure it is in the range of the market. We found that the multiplier is always an important area of discussion and that it needs to be approached with some sense of reasonableness from the parties of the alliance. Just as the contractor cannot work at below-market rates, neither can the owner pay rates higher than the market for work that is performed.

Workload Fluctuations

Probably the greatest challenges for the alliances are the fluctuations that take place in the workload. Even with careful planning, there are ups and downs in workloads. Everyone recognizes that it is important to maintain resources that are trained and experienced in the alliance work practices and specifications. We found that through the systematic rotation of personnel in some of the alliances we were able to build a cadre of trained and experienced resources within Bechtel that could be recalled to the alliances as they were needed.

Appendix 1

Recommended Readings
for Partnering

Author	Book	Publisher
Aguayo, Rafael	Dr. Deming	Simon & Schuster, New York, 1990
Alderson, Wayne T. and Nancy Alderson McDonnell	Theory R Management	Thomas Nelson Publishers, British Columbia, 1994
	Blueprints for Service Quality— The Federal Express Approach	American Management Association, New York, 1991
Bell, Chip R.	Customers as Partners	Berrett-Koehler Publishers, San Francisco, 1994
Bell, Chip R. and Ron Zemke	Managing Knock Your Socks Off Service	American Management Association, New York, 1992
Bhote, Keki R.	Next Operation As Customer (NOAC): How to Improve Quality, Cost and Cycle Time in Service Operations	American Management Association, New York, 1991
Cannie, Joan Koob with Donald Caplin	Keeping Customers for Life	American Management Association, New York, 1991
Champy, James	Reengineering Management	Harper Business, New York, 1995
	Constructability Concepts File	Construction Industry Institute, Publication 3-3, August 1987
	In Search of Partnering Excellence	Construction Industry Institute, Special Publication 17-1, July 1991
Collins, James C. and Jerry I. Porras	Built to Last	Harper Business, New York, 1994
Conway, William E.	The Quality Secret: The Right Way to Manage	Conway Quality, Inc. Nashua, NH 1992

155

Author	Book	Publisher
Daniels, William R.	Orchestrating Powerful Regular Meetings	Pfeiffer & Company, San Diego, 1993
Fisher, Kimball	Leading Self-Directed Work Teams	McGraw-Hill, Inc., New York, 1993
Hammer, Michael and James Champy	Reengineering the Corporation: A Manifesto for Business Revolution	Harper Collins Publishers, Inc., New York, 1993
Hodgetts, Richard M.	Blueprints for Continuous Improvement—Lessons from the Baldrige Winners	American Management Association, New York, 1993
Imlay, John P., Jr.	Jungle Rules: How to be a Tiger in Business	Group Penguin Books, USA, Inc., New York, 1994
Johansson, H.J., and D.K. Carr	Best Practices in Reengineering	McGraw-Hill, Inc., New York, 1995
Johansson, H. J., P. McHugh, A. J. Pendlebury, and W. A. Wheeler	Business Process Reengineering: Breakpoint Strategies for Market Dominance	John Wiley & Sons, Ltd., West Sussex, England, 1993
Keathing, Patrick J. and Stephen F. Jablonsky	Changing Roles of Financial Management—Getting Close to Business	Financial Executive Research Foundation, Morristown, NJ, 1990
Labovitz, George, Yu Sang Chang, Victor Rosansky	Making Quality Work	Oliver Wight Publications, Inc., Essex Junction, VT, 1992
McHugh, Patrick, Giorgio Merli, and William A. Wheeler III	Beyond Business Process Reengineering	John Wiley & Sons, Inc., New York, 1995
Melohn, Tom	The New Partnership	Oliver Wight Publications, Inc., Essex Junction, VT, 1994
Peters, Thomas J. and Robert H. Waterman Jr.	In Search of Excellence	Warner Books, New York, 1982
Thomas, Philip R. and Larry J. Gallace with Kenneth R. Martin	Quality Alone Is Not Enough	American Management Association, New York, 1992
Walton, Mary	The Deming Management Method	Putnam Publishing Group, New York, 1986
Warne, Thomas R.	Partnering for Success	ASCE Press, New York, 1994
Zenger, John H., Ed Musselwhite, Kathleen Hurson, Craig Perrin	Leading Teams	Irwin, New York, 1994
Zimmerman, Larry W. and Glen D. Hart	Value Engineering	Van Nostrand Reinhold Company, New York, 1982

Appendix 2

Partnering Tools

Tool	How Used
1. Objectives/Measures Matrix	A sample of the various metrics used to measure attainment of goals, objectives, and vision.
2. The Deming Principles	The Deming management principles.
3. Development of Project Goals	Flow chart of the development of project goals process.
4. The Quality Improvement Process	A flow chart of the Quality Improvement Process.
5. The Continuous Improvement Cycle	The Deming Continuous Improvement Cycle.
6. The Spirit of Involvement Cycle	The cycle which shows how partnering causes self-actualization in people.
7. Typical Project Task Force	Shows a typical project task force organization chart
8. Simplified Project Execution Flow chart	——
9. Project Management Feedback and Self-Evaluation Form	Used before project management meetings to assess the team members' evaluation of partnering principles.

Tool	How Used
10. Field Working Group Project Evaluation Form	Used before project management meetings to assess the field workinggroups' evaluation of the partnering on the project.
11. Project Action Item List	To keep a numerical listing of those items that have been delegated as problems and when they were completed.
12. Partnering versus Partnerships versus Alliances	A table of the similarities and differences of partnering, partnerships, and alliances.
13. U.S. Army Corps of Engineers Partnering Contract Clause	A copy of a typical USACE partnering contract clause.
14. Design Project Phase Definitions	Project phase definitions for a design construction project.

OBJECTIVES	TECHNICAL SURVEY	CUSTOMER SURVEY	SAFETY METRICS	PROJECT METRICS	EMPLOYEE SURVEY
1. Meet Customer Expectations	•	•		•	
2. Challenge Each Other	•	•			•
3. Reduce TIC by 25% from Historic Norms				•	
4. Reduce Project Cycle Time				•	
5. Provide Reliable, Superior Technical Resources to Support Business and Plant Requirements	•	•		•	•
6. Provide Resources and Work Processes that are Complementary and Cost Effective	•	•		•	•
7. Perform Work Without any Safety or Environmental Incidents	•	•	•		•
8. Utilize "Best Practices"				•	
9. Apply Innovation and Creativity	•	•		•	
10. Measure Results	•	•	•	•	•
11. Demonstrate Continuous Improvement		•		•	•
12. Promote Teamwork, Trust and Open Communications	•	•			•
13. Foster Personally Rewarding Work Environment					•
14. Pay for Performance by Sharing Risks / Rewards.		•			

Figure A2.1. *Objective/measures matrix.*

THE DEMING PRINCIPLES

The Fourteen Points

1. Create constancy of purpose for improvement of product and service.
2. Adopt quality as a religion.
3. Cease dependence on mass inspection.
4. End the practice of awarding business on price tag alone.
5. Improve constantly and forever the system of production and service.
6. Institute training.
7. Institute leadership.
8. Drive out fear.
9. Break down barriers between staff areas.
10. Eliminate slogans, exhortations, and targets for the workhorse.
11. Eliminate numerical quotas.
12. Remove barriers to pride of workmanship.
13. Institute a vigorous program of education and retraining.
14. Take action to accomplish the transformation, the quality mission.

The Seven Deadly Diseases

1. Lack of constancy of purpose
2. Emphasis on short-term profits
3. Evaluation by performance, merit ratings, or annual review
4. Mobility of management
5. Running a company on visible figures alone
6. Excessive medical costs
7. Excessive costs of warranty, fueled by lawyers that work on contingency fees

Obstacles that Thwart Productivity

1. Neglect of long-range planning
2. Relying on technology to solve problems
3. Seeking examples to flow rather than developing solutions
4. Making excuses such as "our problems are different"

Figure A2.2.

DEVELOPMENT OF PROJECT GOALS

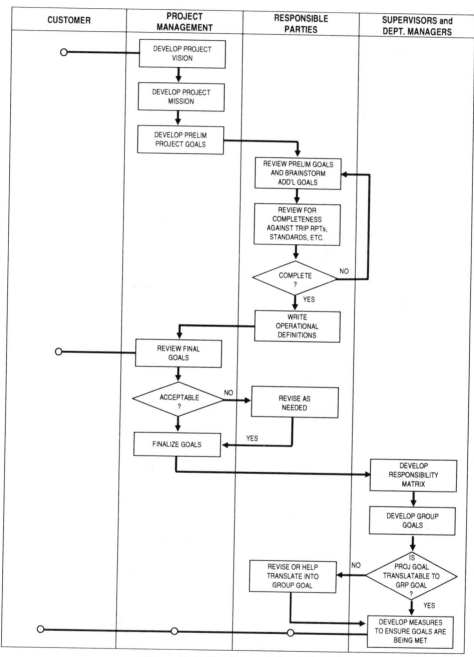

Figure A2.3. *Development of project goals.*

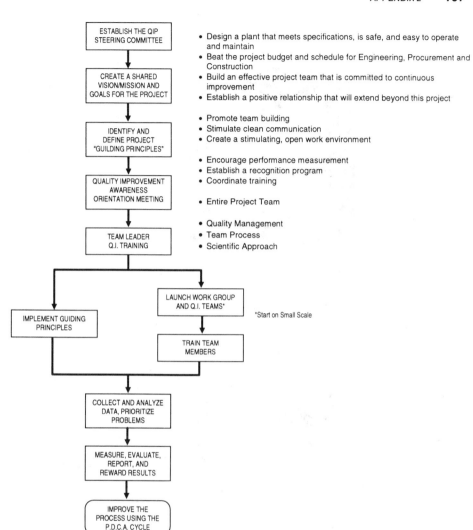

Figure A2.4. *The quality improvement process.*

ADAPT TO CHANGE

PLAN THE OPPORTUNITY

CHECK THE RESULTS

DO THE CHANGE

Figure A2.5. *The continuous improvement cycle.*

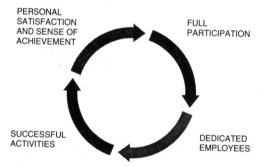

PERSONAL SATISFACTION AND SENSE OF ACHIEVEMENT

FULL PARTICIPATION

SUCCESSFUL ACTIVITIES

DEDICATED EMPLOYEES

Figure A2.6. *The spirit of involvement cycle.*

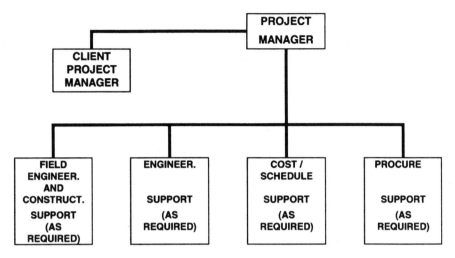

Figure A2.7. Typical project task force.

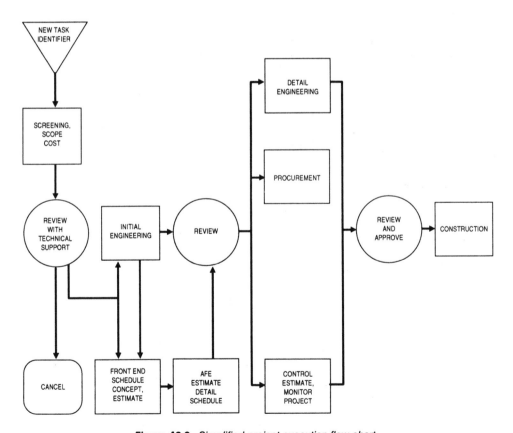

Figure A2.8. Simplified project execution flow chart.

The Purpose of this form is to assess the health and performance of the project management team. As a stakeholder in the project, your honest and open evaluation is essential to the success of the team in achieving its stated goals. Your input will be reviewed and discussed as part of the project management meeting agenda. Please complete this form and send a copy to _____(Fax _____) no later than two days prior to the scheduled project management meeting. (NR = Not Rated)

1. **Responsiveness:**
The ability to react to issues and come to closure in a timely manner.

1	2	3	4	5	NR
No obvious special effort (Status quo relative to other projects).		Effort shown but erratic.		Obvious high priority given to closing essentially every issue.	

2. **Commitment:**
The demonstrated ability to live up to their agreement to manage this project as a Partner Alliance.

1	2	3	4	5	NR
Individuals have retreated to traditional roles.		Only seen when stakes are low.		Every member bound to the agreement.	

3. **Information Flow:**
The ability to exchange/ share information among members.

1	2	3	4	5	NR
Critical information closely held. Apparent hidden agendas.		Shared but can be more open with trust.		All information freely and openly shared.	

4. **Cooperation:**
The ability to work as a team towards mutual goals and objectives.

1	2	3	4	5	NR
No more than non partnered projects. Traditional roles.		Sense of teamwork varies. Cooperation among some members.		All members willing and demonstrate their ability to work together.	

5. **Communications:**
The ability to communicate either verbally or through written or established processes in a clear and effective manner.

1	2	3	4	5	NR
Only traditional communication processes used.		Open but with some misunderstanding.		Open and free flowing; clear and unencumbered.	

Figure A2.9. *Project management feedback and self-evaluation form.*

6. Trust:
The ability to rely on the character, ability, strengths, and truth of the individual members.

1	2	3	4	5	NR
Traditional roles with reliance on legal ground only.		Only when the stakes are low. Confidence moderate or lightly placed.		Fully placed with members demonstrating confidence.	

7. **Performance:**
The ability to execute and complete/accomplish tasks undertaken.

1	2	3	4	5	NR
Tasks completed by individuals unilaterally or with extreme effort as a group.		Tasks dealt with as a group with mixed results.		Tasks effectively carried out and followed through. If agreed to, it's done.	

8. **Process Effectiveness:**
The ability to manage project processes. (Submittals/Mods/ etc.)

1	2	3	4	5	NR
Processes bureaucratic and slow with little quality added. Status quo.		Processes in place but are in need of refinement.		Processes understood, followed, monitored and efficient. No project delays observed.	

9. **Problem Resolution:**
The ability to resolve problems.

1	2	3	4	5	NR
Decisions made by higher levels with unilateral authority.		Consensus is sometimes reached or authority held too high.		Issues are dealt with not personalities; consensus is reached and decisions made at the lowest appropriate level.	

10. **Downward Partnering:**
The ability to ensure all members of the team are effectively partnered.

1	2	3	4	5	NR
Team members not working together, mistrust, miscommunication and reliance on traditional roles.		Team members working together but maintenance needed.		Group working as a cohesive team with trust and open communications. Partnering goals met.	

11. **Early Warning System**
The ability to identify threats to either the Partnering Alliance or project schedule and/ or cost.

1	2	3	4	5	NR
Team reactive with little or unheeded warning.		Most threats identified early but response needs improvement.		Potential threats readily discussed and when identified, dealt with minimum impact to cost/ schedule.	

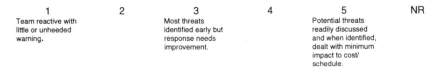

Figure A2.9. (Continued)

12. Top 5 Issues:
Please list your top five issues / concerns for the project.

1._____

2._____

3._____

4._____

5._____

Name: _____

Date: _____

Figure A2.9. (Continued)

Project Management Survey Results
Period of: _____

Question	Ave.									
1. Response										
2. Commitment										
3. Information Flow										
4. Cooperation										
5. Communication										
6. Trust										
7. Performance										
8. Process Effectiveness										
9. Problem Resolution										
10. Downward Partnering										
11. Early Warning										

Top 5 Issues

1										
2										
3										
4										
5										

Figure A2.9. (Continued)

The purpose of this form is to asses the health and performance of the project. As a stakeholder, your honest and open evaluation is essential to the success of the team in achieving its stated goals. Your input will be reviewed as part of the next management meeting agenda. Please complete this form and give it to your management representative. (NR= Not Rated)

1. Quality:
The project is being constructed with quality in mind.

1	2	3	4	5	NR
Quality is not being considered.		Effort shown but erratic.		Obvious high priority given to quality issues.	

2. Active Participation:
All team members are contributing to the project.

1	2	3	4	5	NR
Hardly anyone seems interested.		Some seem to not want to get involved.		Excellent participation by all parties.	

3. Cooperation:
The Ability to work as a team toward mutual goals and objectives.

1	2	3	4	5	NR
No more than non-partnered projects. Traditional roles.		Sense of teamwork varies. Cooperation among some members.		All members willing and demonstrate their ability to work together.	

4. Feedback:
Information coming back to you on project status, problems resolution, etc.

1	2	3	4	5	NR
Little information is being communicated.		Open but requires follow-up efforts to get information.		Open and free-flowing. Processes clear and unencumbered.	

5. Project Management:
How the project is being managed by the management group.

1	2	3	4	5	NR
Apparent lack of control.		Project control good, but could be better.		Fully in control.	

Figure A2.10. Field working group project evaluation form.

6. Safety:

The project is being conducted with the proper emphasis on safety.

1	2	3	4	5	NR
Safety measures not being considered.		Safety controls O.K., but needs more work.		Safety awareness evident.	

7. Top 3 Issues:

Please list your top three issues/ concerns for the project:

1. _____
   _____
   _____

2. _____
   _____
   _____

3. _____
   _____
   _____
   _____
   _____

My organization is:

Contractor   PM   Engineer   Design   Sub   COE

Figure A2.10. (Continued)

Field Work Group
Survey Results
Period of: _____

Question Number	Question Topic	Overall Average	Contractor	PM	Engineer	Design	Sub	COE
1.	Quality							
2.	Active Participation							
3.	Cooperation							
4.	Feedback							
5.	Project Management							
6.	Safety							

ISSUES:

1.

2.

3.

Figure A2.10. (Continued)

Page: _____ of _____
Date Updated: _____

Item No.	Action Item	Responsible Individual	Date Opened	Forecast Complete	Actual Complete	Status	Priority

Priority : 1= Immediate
          2= Routine

Figure A2.11. *Project action item list.*

	Partnering Concept		Partnership	Alliance
Basic Concept:	The pursuit of a long-term base business activity established, usually under a formal document, with a "culturally" acceptable entity. Can be shorter term and involve more than two parties - e.g. engineer, contractor and owner for one project. Partnering is a "soft" management method using team building concepts to overcome cultural parameters and focus the parties' attention and efforts on achievement of some agreed to goal/mission/vision.		A formal business arrangement between parties to jointly pursue a business or activity. General partners are liable to the full extent of their assets.	A formal business arrangement between parties to jointly pursue a business or activity.
	Contract	**Partnering Method**		
Documentation	A formal document dealing with the formal arrangement such as a EPC contract for services. [Should mention "partnering" methods / techniques.]	A written goal/vision/mission statement. Sometimes signed by the individuals on the team committing them. Although formal, not intended to modify contract.	Formal contract setting out the financial, legal and organizational parameters for the partnership.	Formal contract setting out the financial, legal and organizational parameters for the partnership.
Risk sharing	Risks formally allocated in the contract. [Usually deals with professional services rendered. May be limited.]	Sharing of informal risks, particularly those attributed to the performance or achievement(s) of the "team."	Risks detailed in contract usually deal with financial and legal risks of business activity. May cover professional service risks. General partners are exposed to the full extent of their assets.	Contract usually establishes separate liability and disclaims partnership.
Contributions	No direct payment for partnering. [Usually a performance of professional services that are then reimbursed.] See documentation above.	Conciliation and concession to "team" requirements.	Usually contribution of some assets or cash to provide ability to fund partnership organization in achieving goals of business for profit.	Usually separate with limited pooling.
Compensation	Compensation for services performed. Performance provides base workload. Some provision for incentive or reward sharing if achievement of some goal. Amounts predetermined or pre limited.	Usually no separate compensation arrangements.	Share in the profits of the partnership as per formal contract arrangement. Can be payment for some services, but profits are shared on basis of provisions in contract.	Usually separate payments without sharing.
Organization boundaries	Different legal entities working at arm's length, usually not mentioned.	Attempt to overcome the organizational boundaries by focusing on team building to achieve goal / vision. Team members feel a high degree of self actualization and achievement. They make a noticeable difference.	Creation of new legal entity.	Creation of a business relationship which is not a legal entity.
Term	Usually concerned with a definite period. Usually for some specified period, but may be on a continuing basis.	Partnering can be "evergreen" if it continues to satisfy the needs of all the parties.	Usually for the achievement of some specific long-term business objective.	Usually for a specific business objective, usually shorter term than partnership.
Measurement	Usually some physical or financial result. [Performance against a budget or forecast.]	Working process and goals focused on increasing or enhancing value. Measurement of process achievement, can be tangible, or intangible.	Financial statements reflecting financial results of the partnership as a separate legal entity.	Alliance parties usually have separate accounting.

TYPICAL U.S. ARMY CORPS OF ENGINEERS PARTNERING CONTRACT CLAUSE

PARTNERING: As described in the Project Management and Partnership Plan, the Government proposes to form a cohesive Partnership Agreement with the A-E and its subcontractors. This partnership will strive to draw on the strengths of each organization in an effort to achieve a quality design, on schedule and within budget. The Partnership Agreement shall be made and maintained through the use of professionally facilitated workshops and nurturing sessions, will be bilateral and totally voluntary. Any costs associated with partnering activities will be agreed upon by both parties and will be shared equally with no change in contract price.

Note

1. The word Partnership is used by the government although it has a special legal meaning.
2. Partnering is voluntary.
3. The costs of "partnering" are shared equally.

Figure A2.13

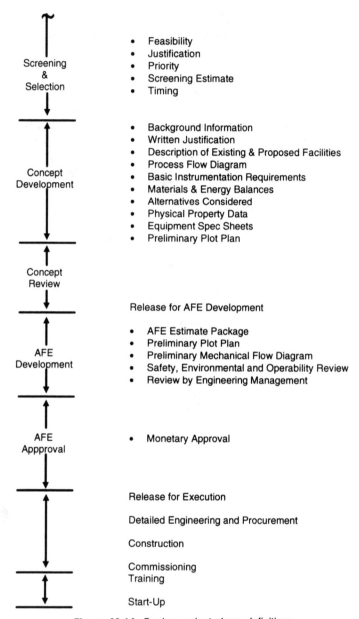

Figure A2.14. *Design project phase definitions.*

About the Authors

Henry J. Schultzel, CPA, CMC, is partner in charge of the engineering and construction consulting practice of Coopers & Lybrand L.L.P. and serves as the chairman of the engineering and construction industry program for the firm.

For over 27 years, Mr. Schultzel has worked extensively with many of the major engineering and construction firms, directing numerous large-scale engagements both in the United States and abroad. He has successfully implemented the concepts of partnering in assisting clients to achieve optimal performance in industries covering transportation systems, including rail, mass transit, airports, and seaports; municipal utilities; power and energy, including fossil, nuclear, and geothermal; and significant heavy manufacturing projects involving a variety of industry applications.

Mr. Schultzel studied electrical engineering at Fairleigh Dickinson University in 1961, received a Bachelor of Science degree in accounting and business administration from Saint Peter's College in 1966, and attended the Seton Hall University of Law in 1968. Mr. Schultzel also served as the Commander of Special Forces (5th Group SOG) in the Pleiku Region of the Republic of South Vietnam. He is a licensed certified public accountant in California, New York, and Louisiana and a member of the American Institute of Certified Public Accountants, the California State Society of Certified Public Accountants, and New York State Society of Certified Public Accountants, and the Institute of Management Consultants.

Born in New Jersey, Hank joined the New York office of Coopers & Lybrand in 1967 before transferring to the San Francisco office in 1973. He and his wife, Ann, live in Danville with their twin sons, Matthew and Mark, the youngest of their five children.

V. Paul Unruh, CPA, is a senior vice president, chief financial officer, and treasurer of the Bechtel Group, Inc., the Bechtel Capital Management Corporation, and the principal operating companies of the Bechtel Group.

Since joining Bechtel in 1978, Mr. Unruh has served in a leadership role in several key initiatives supporting Bechtel's fundamental business strategies. Most notably, he conceived and implemented a network of highly automated, integrated financial applications that today link Bechtel's worldwide financial operations. He had led a successful major restructuring of Bechtel's financial operations and has been a proponent of organizational and management changes that have helped to make Bechtel one of the world's preeminent engineers and constructors.

Mr. Unruh received Bachelor of Science and Master of Science degrees in accounting from the University of North Dakota and served as an officer in the U.S. Army Finance Corps. He is a licensed certified public accountant in California and North Dakota and is a member of the Financial Executives Institute, Association for Systems Management, Society of International Treasurers, American Institute of Certified Public Accountants, and the California and North Dakota Societies of Certified Public Accountants.

Prior to joining Bechtel, Mr. Unruh practiced public accounting with Coopers & Lybrand. He and his wife, Eileen, live with their two children in Orinda, California.

Index